Adversarial Machine Learning

Synthesis Lectures on Artificial Intelligence and Machine Learning

Editors
Ronald J. Brachman, *Jacobs Technion–Cornell Institute at Cornell Tech*
Peter Stone, *University of Texas at Austin*

iv

Adversarial Machine Learning

Yevgeniy Vorobeychik and Murat Kantarcioglu

ISBN: 978-3-031-00452-0   paperback
ISBN: 978-3-031-01580-9   ebook
ISBN: 978-3-031-02708-6   epub
ISBN: 978-3-031-00025-6   hardcover

DOI 10.1007/978-3-031-01580-9

A Publication in the Springer series
SYNTHESIS LECTURES ON ARTIFICIAL INTELLIGENCE AND MACHINE LEARNING
Lecture #38
Series Editors: Ronald J. Brachman, *Jacobs Technion–Cornell Institute at Cornell Tech*
            Peter Stone, *University of Texas at Austin*
Series ISSN
Print 1939-4608   Electronic 1939-4616

Adversarial Machine Learning

Yevgeniy Vorobeychik
Vanderbilt University

Murat Kantarcioglu
University of Texas, Dallas

SYNTHESIS LECTURES ON ARTIFICIAL INTELLIGENCE AND MACHINE LEARNING #38

ABSTRACT

The increasing abundance of large high-quality datasets, combined with significant technical advances over the last several decades have made machine learning into a major tool employed across a broad array of tasks including vision, language, finance, and security. However, success has been accompanied with important new challenges: many applications of machine learning are adversarial in nature. Some are adversarial because they are safety critical, such as autonomous driving. An adversary in these applications can be a malicious party aimed at causing congestion or accidents, or may even model unusual situations that expose vulnerabilities in the prediction engine. Other applications are adversarial because their task and/or the data they use are. For example, an important class of problems in security involves detection, such as malware, spam, and intrusion detection. The use of machine learning for detecting malicious entities creates an incentive among adversaries to evade detection by changing their behavior or the content of malicius objects they develop.

The field of adversarial machine learning has emerged to study vulnerabilities of machine learning approaches in adversarial settings and to develop techniques to make learning robust to adversarial manipulation. This book provides a technical overview of this field. After reviewing machine learning concepts and approaches, as well as common use cases of these in adversarial settings, we present a general categorization of attacks on machine learning. We then address two major categories of attacks and associated defenses: decision-time attacks, in which an adversary changes the nature of instances seen by a learned model at the time of prediction in order to cause errors, and poisoning or training time attacks, in which the actual training dataset is maliciously modified. In our final chapter devoted to technical content, we discuss recent techniques for attacks on deep learning, as well as approaches for improving robustness of deep neural networks. We conclude with a discussion of several important issues in the area of adversarial learning that in our view warrant further research.

Given the increasing interest in the area of adversarial machine learning, we hope this book provides readers with the tools necessary to successfully engage in research and practice of machine learning in adversarial settings.

KEYWORDS

adversarial machine learning, game theory, machine learning

Contents

List of Figures

Preface

The research area of adversarial machine learning has received a great deal of attention in recent years, with much of this attention devoted to a phenomenon called *adversarial examples*. In its common form, an adversarial example takes an image and adds a small amount of distortion, often invisible to a human observer, which changes the predicted label ascribed to the image (such as predicting gibbon instead of panda, to use the most famous example of this). Our book, however, is not exactly an exploration of adversarial examples. Rather, our goal is to explain the field of adversarial machine learning far more broadly, considering supervised and unsupervised learning, as well as attacks on training data (poisoning attacks) and attacks at decision (prediction) time, of which adversarial examples are a special case. We attempt to convey foundational concepts in this rapidly evolving field, as well as technical and conceptual advances. In particular, the flow of the book, beyond introductory materials, is to describe algorithmic techniques used in attacking machine learning, followed by algorithmic advances in making machine learning robust to such attacks. Nevertheless, in the penultimate chapter we provide an overview of some of the recent advances specific to the deep learning methods. While it is important to see such methods within the broader area of adversarial learning, the motivation, techniques, and empirical observations documented in this last chapter are most salient in the context of deep neural networks (although many of the technical approaches are in principle quite general).

This book assumes a great deal from the reader. While there is an introduction to machine learning concepts, terminology, and notations, some level of prior familiarity with machine learning is likely needed to fully grasp the technical content. Additionally, we expect a certain degree of maturity with statistics and linear algebra, and some prior knowledge of optimization (in particular, remarks about convex optimization, and discussions of techniques such as gradient descent, assume familiarity with such concepts).

Yevgeniy Vorobeychik and Murat Kantarcioglu
June 2018

Acknowledgments

We wish to acknowledge the many colleagues and students who helped bring this book to life, either by engaging with us in related research, or by commenting on some of the content either in written or presented form, and correcting errors. In particular, we thank Bo Li, Chang Liu, Aline Oprea for their contributions to some of the technical content and numerous related discussions. We are also indebted to a number of people for discussions on the topics presented in this book, including Daniel Lowd, Pedro Domingos, Dawn Song, Patrick McDaniels, Milind Tambe, Arunesh Sinha, and Michael Wellman. We are especially grateful to Matthew Sedam for identifying a number of errors in the presentation, and to Scott Alfeld and Battista Biggio for their suggestions that significantly improved the manuscript. Finally, we gratefully acknowledge the funding sources which enabled both this book, and many related research articles: the National Science Foundation (grant IIS-1649972), Army Research Office (grant W911NF-16-1-0069), Office of Naval Research (grant N00014-15-1-2621), and the National Institutes of Health (grant R01HG006844).

Yevgeniy Vorobeychik and Murat Kantarcioglu
June 2018

CHAPTER 1

Introduction

As machine learning techniques have entered computing mainstream, their uses have multiplied. Online advertising and algorithmic trading are now inconceivable without machine learning, and machine learning techniques are increasingly finding their ways into health informatics, fraud detection, computer vision, machine translation, and natural language understanding. Of most importance to this book, however, is the increasing application that machine learning techniques are finding in security in general, and cybersecurity in particular. The reason is that security problems are, by definition, adversarial. There are the *defenders*—the good guys—for example, network administrators, anti-virus companies, firewall manufacturers, computer users, and the like, trying to maintain productivity despite external threats, and *attackers*—the bad guys—who spread malware, send spam and phishing emails, hack into vulnerable computing devices, steal data, or execute denial-of-service attacks, for whatever malicious ends they may have.

A natural role for machine learning techniques in security applications is detection, examples of which include spam, malware, intrusion, and anomaly detection. Take detection of malicious email (spam or phishing) as a prototypical example. We may start by obtaining a labeled dataset of benign and malicious (e.g., spam) emails, containing the email text and any other relevant information (for example, using metadata such as the DNS registration information for the sender IP). For illustration, let's focus on email text as sole information about the nature (malicious or benign) of the email. The dataset is transformed into feature vectors which capture the text content, and numerical labels corresponding to the two classes (malicious vs. benign). A common way to numerically represent a document is by using a *bag-of-words* representation. In a bag-of-words representation, we construct a dictionary of words which may appear in emails, and then create a feature vector for a given email by considering how often each word in the dictionary has appeared in the email text. In the simpler binary bag-of-words representation, each feature simply indicates whether the corresponding word has appeared in the email text; an alternative real-valued representation considers the number of times a word appears in the email, or term frequency-inverse document frequency (tf-idf) [Rajaraman and Ullman, 2012]. Once the dataset is encoded into a numeric format, we train a classifier to predict whether a new email is spam or phish based on the email text in its bag-of-words feature representation.

For a large enough dataset, using state-of-the-art machine learning (classification) tools can enable extremely effective spam or phishing email detection, *with respect to an evaluation using past data*. What makes this setting *adversarial* is that spam and phishing emails are generated deliberately, with specific goals in mind, by malicious actors. Said actors are decidedly unhappy if their emails are detected and as a result fail to reach the intended destination (i.e., user mailboxes). The choice of the spammers is two-fold: either get out of the business of sending spam, or change the way they construct the spam templates so as to *evade* spam detectors. Such adversarial classifier evasion by spammers is a prototypical use case of adversarial machine learning.

Figure 1.1 illustrates spam/phishing detector evasion through an example. In this case, a malicious party had crafted a phishing email, shown in Figure 1.1 (left), which attempts to deceive recipients into clicking on an embedded malicious link (the embedded link itself is stripped in this illustration). Now, suppose that an effective spam detector is deployed, such that the email on the left is categorized as malicious and filtered by the spam filter. The originator of the phishing email can rewrite the text of the email, as shown in the example in Figure 1.1 (right). Considering the two emails side-by-side demonstrates the nature of the attack. On the

Greetings,

After reviewing your Linkedin profile, our company would like to present you a part-time job offer as a finance officer in your region. This job does not require any previous experience. Here is a list of tasks that our employee should accomplish:
1. Receive payment from our customers into your bank account.
2. Keep your commission fee of 10% from the payment amount.
3. Send the rest of the payment to one of our payment receivers in Europe via Moneygram or Western Union.

For more details of the job, click here.
After enrolling to our-part time job you will be contacted by one of our human resource staff.

Thanks,
Karen Hoffman.
Human Resource Manager.

Greetings,

Our company is looking to expand and after reviewing your Linkedin profile, we would like to present you a part-time job offer as a finance officer in your region. This job does not require any previous experience.

For more details about this job offer, click here.

After enrollment you will be contacted by one of our human resource staff.

Thanks,
Karen Hoffman,
Human Resource Manager.

Figure 1.1: Example of adversarial evasion in the context of phishing. Left: example original phishing email; right: a reworded phishing email that is classified as benign.

one hand, the general message content still gets the primary point across, which communicates the possibility of easy financial gain to the recipient, and still guides them to click on the embedded malicious link. On the other hand, the message text is now sufficiently different from the original as to no longer appear malicious to the spam detector learned on past data. This is typically accomplished by removing "spammy" words (that is, words that tend to increase the maliciousness score of the detector) and potentially adding words which the detector treats as benign (the latter strategy is often called the *good-word attack* [Lowd and Meek, 2005b]).

These two conflicting goals—evading detection while at the same time achieving original attack goals—are central to evasion attacks in general. In malware detection, as in spam, the attacker would wish to modify malware code to appear benign to a detector, while maintaining the original or equivalent malicious functionality. In intrusion detection systems, a hacker would wish to revise the automated tools and manual procedures so as to appear benign, but at the same time still successfully execute the exploit.

While evasion attacks seem most natural in adversarial uses of machine learning, numerous other examples illustrate that the scope of the adversarial machine learning problem is far broader. For example, obtaining labeled training data with the purpose of learning detectors exposes learning algorithms to *poisoning* attacks, whereby malicious actors manipulate data which is subsequently used to train learning algorithms. Forensic analysis of malware or hacking attempts may wish to use clustering in an attempt to categorize the nature of the attack, including its attribution, but doing so may be susceptible to deliberate attacks which manipulate cluster assignment by slightly changing the nature of attacks, causing mistaken categorization and attribution. Moreover, adversarial learning also has broader scope than cybersecurity applications. In physical security, for example, the problem of detecting malicious activity through video surveillance is adversarial in nature: clever attackers may manipulate their appearance, or other factors such as how they go about their malicious activity, to avoid detection. Similarly, credit card fraud detection, which uses anomaly detection methods to determine whether a particular activity is alarming due to its highly unexpected nature, may be susceptible to attacks which make transactions appear typical for most credit card users. As yet another example, algorithmic trading techniques which use machine learning may be susceptible to exploitation by competitors who make market transactions with the sole purpose of manipulating *predicted* prices, and using the resulting arbitrage opportunities to make a profit (these are commonly known as spoofing orders [Montgomery, 2016]).

Systematic study of *adversarial machine learning* aims to formally investigate problems introduced by the use of machine learning techniques in adversarial environments in which an intelligent adversary attempts to exploit weaknesses in such techniques. The two general aspects of this research endeavor are: (1) modeling and investigation of attacks on machine learning and (2) developing learning techniques which are robust to adversarial manipulation.

In this book, we study a number of common problems in adversarial learning. We start with an overview of standard machine learning approaches, with a discussion of how they can be applied in adversarial settings (Chapter 2); this chapter allows us to fix notation, and provides background to the core content of the book. Next, we present a categorization of attacks on machine learning methods to provide a general, if not fully comprehensive, conceptual framework for the detailed presentation which follows (Chapter 3). We then consider a problem of attacks on decisions made by learning models (Chapter 4), and subsequently proceed to discuss a number of techniques for making learning algorithms robust to such attacks (Chapter 5). Thereafter, we consider the problem of poisoning the training data used by learning algorithms (Chapter 6), followed by a discussion of techniques for making algorithms robust to poisoned training data (Chapter 7). The final chapter of this book (Chapter 8) addresses a more recent variation of the adversarial learning specifically dealing with deep neural networks in computer vision problems. In this chapter, we provide an overview of the major classes of attacks on deep learning models for computer vision, and present several approaches for learning more robust deep learning models.

CHAPTER 2

Machine Learning Preliminaries

To keep this book reasonably self-contained, we start with some machine learning basics. Machine learning is often broadly divided into three major areas: supervised learning, unsupervised learning, and reinforcement learning. While in practice these divisions are not always clean, they provide a good point of departure for our purposes.

We start by offering a schematic representation of learning, shown in Figure 2.1. In this schematic representation, learning is viewed as a pipeline which starts with raw data, for example, a collection of executable files, with associated labels indicating whether a file is benign or malicious. This raw data is then processed to extract numeric features from each instance i, obtaining an associated feature vector x_i (for example, this could be a collection of binary variables indicating presence in an executable of particular system calls). This becomes *processed data*, but henceforth we call it simply *data*, as it is to this processed dataset that we can apply learning algorithms—the next step in the pipeline. Finally, the learning algorithm outputs a *model*, which may be a mathematical model of the data (such as its distribution) or a function that predicts labels on future instances.

Figure 2.1: A schematic view of machine learning.

2.1 SUPERVISED LEARNING

In *supervised learning*, you are given a model class $\mathcal{F}$ and a dataset $\mathcal{D} = \{x_i, y_i\}_{i=1}^n$ of feature vectors $x_i \in \mathcal{X} \subseteq \mathbb{R}^m$, where $\mathcal{X}$ is the *feature space*, and labels y_i from some label set $\mathcal{Y}$. This dataset is typically assumed to be generated i.i.d. from an unknown distribution $\mathcal{P}$, i.e., $(x_i, y_i) \sim \mathcal{P}$. The ultimate goal (the "holy grail") is to find a model $f \in \mathcal{F}$ with the property that

$$\mathbb{E}_{(x,y)\sim\mathcal{P}}[l(f(x), y)] \leq \mathbb{E}_{(x,y)\sim\mathcal{P}}[l(f'(x), y)] \quad \forall f' \in \mathcal{F}, \tag{2.1}$$

where $l(f(x), y)$, commonly called the *loss function*, measures the error that $f(x)$ makes in predicting the true label y. In simple terms, if there is some "true" function h we are trying to learn, the goal is to find $f \in \mathcal{F}$ which is as close to h as possible, given the constraints that the model class imposes on us. Having this notion of a target function h in mind, we can restate (2.1) as

$$\mathbb{E}_{x \sim \mathcal{P}}[l(f(x), h(x))] \leq \mathbb{E}_{x \sim \mathcal{P}}[l(f'(x), h(x))] \quad \forall \ f' \in \mathcal{F}. \tag{2.2}$$

This special case may be somewhat easier to have in mind moving forward.

In practice, since $\mathcal{P}$ is unknown, we use the training data $\mathcal{D}$ in order to find a candidate f which is a good approximation of the labels actually observed in this data. This gives rise to the following problem of minimizing *empirical risk* (commonly known as empirical risk minimization, or ERM):

$$\min_{f \in \mathcal{F}} \sum_{i \in \mathcal{D}} l(f(x_i), y_i) + \gamma \rho(f), \tag{2.3}$$

where it is common to add a regularization term $\rho(f)$ which penalizes the complexity of candidate models f (in the spirit of Occam's razor that among multiple equally good models one should favor the simpler one). Commonly, functions in the model class $\mathcal{F}$ have a parametric representation, with parameters w in a real vector space. In this case, we typically write the ERM problem as

$$\min_{w} \sum_{i \in \mathcal{D}} l(f(x_i; w), y_i) + \gamma \rho(w), \tag{2.4}$$

with regularization $\rho(w)$ often taking the form of an l_p norm of w, $\|w\|_p^p$; common examples include l_1 norm, or lasso, $\|w\|_1$, and l_2 norm, $\|w\|_2^2$.

Supervised learning is typically subdivided into two categories: *regression*, where labels are real-valued, i.e., $\mathcal{Y} = \mathbb{R}$, and *classification*, where $\mathcal{Y}$ is a finite set of labels. We briefly discuss these next.

2.1.1   REGRESSION LEARNING

In regression learning, since the labels are real values, it is unlikely we will ever get them *exactly* right. An appropriate loss function would penalize us for making predictions that are far from the observed labels. Typically, this is captured by an l_p norm:

$$l(f(x), y) = \|f(x) - y\|_p^p.$$

Indeed, it is rarely useful to consider anything other than the l_1 norm or the l_2 (Euclidean) norm for this purpose.

To make regression learning more concrete, consider linear regression as an example. In this case, the model class $\mathcal{F}$ is the set of all linear functions of dimension m, or, equivalently, the set of all coefficients $w \in \mathbb{R}^m$ and an offset or bias term, $b \in \mathbb{R}$, with an arbitrary linear function being $f(x) = w^T x + b = \sum_{j=1}^{m} w_j x_j + b$. If we introduce an additional constant feature

$x_{m+1} = 1$, we can equivalently write the linear model as $f(x) = w^T x$; henceforth, we often use the latter version. Our goal is to find some parametrization w to minimize error on training data:

$$\min_{w \in \mathbb{R}^m} \sum_{i \in \mathcal{D}} l(w^T x_i, y_i).$$

The commonly used standard ordinary least squares (OLS) regression allows us to solve this problem in closed form by setting the first derivative to zero using Euclidean (l_2) norm as the loss function. Imposing l_1 (lasso) regularization (i.e., adding $\|w\|_1$ to the objective) will typically result in a *sparse* model, where many of the feature weights $w_j = 0$. With l_2 regularization ($\|w\|_2^2$), on the other hand (resulting in *ridge regression*), model coefficients shrink in magnitude, but will in general not be exactly 0.

2.1.2   CLASSIFICATION LEARNING

In the most basic version of classification learning we have two classes. One convenient way to encode these is $\mathcal{Y} = \{-1, +1\}$, and this encoding is particularly natural in adversarial settings, where -1 means "benign" (normal emails, legitimate network traffic, etc.) while $+1$ corresponds to "malicious" (spam, malware, intrusion attempt). This is convenient because we can now represent a classifier in the following form:

$$f(x) = \text{sgn}\{g(x)\},$$

where $g(x)$ returns a real value. In other words, when $g(x)$ is negative, we return -1 as the class, while a positive $g(x)$ implies that $+1$ is returned. *This decomposition will be crucial to understand the nature of many attacks on classification.* We will call $g(x)$ a *classification score function*, or simply a score function.

In classification, the "ideal" loss function is typically an indicator function which is 1 if the predicted label does not match the actual, and 0 otherwise (this is commonly known as the 0/1 loss). The main challenge with this loss function is that it is non-convex, making the empirical risk minimization problem quite challenging. As a consequence, a number of alternatives are commonly used instead.

One general approach is to use a *score-based* loss function which takes the score function $g(x)$ rather than $f(x)$ as input. To see how such a loss function can be constructed, observe that a classification decision is correct (i.e., $f(x) = y$) iff

$$yg(x) \geq 0,$$

that is, both y and $g(x)$ have the same sign. Moreover, whenever their signs differ, a larger $|yg(x)|$ means that this difference is larger—in other words, a score function allows us to also assign a *magnitude* of error. Thus, score-based loss functions are naturally represented by $l(yg(x))$.

As an example, the 0/1 loss becomes

$$l_{01}(yg(x)) = \begin{cases} 1 & \text{if} \quad yg(x) \geq 0 \\ 0 & \text{o.w.} \end{cases}$$

This, of course, remains non-convex. A typical approach is to use a convex relaxation of the 0/1 loss function in its place, resulting, for a $g(x)$ which is convex in model parameters, in a convex optimization problem. Common examples are *hinge loss*, $l_h(yg(x)) = \max\{0, 1 - yg(x)\}$ (used by *support vector machines*), and *logistic loss*, $l_l(yg(x)) = \log(1 + e^{-yg(x)})$ (used by a logistic regression).

As an illustration, consider linear classification. In this case, $g(x) = w^T x$ is a linear function, and, just as in linear regression, we aim to find an optimal vector of feature weights w. If we use the hinge loss along with l_2 regularization, we obtain the following optimization problem for ERM:

$$\min_w \sum_{i \in D} \max\{0, 1 - y_i w^T x_i\} + \gamma \|w\|_2^2,$$

which is just the optimization problem solved by the linear support vector machine [Bishop, 2011].

One way to generalize these ideas to multi-class classification problems is to define a general scoring function $g(x, y)$ for feature vector x and class label y. Classification decision is then

$$f(x) = \arg \max_{y \in \mathcal{Y}} g(x, y).$$

As an example, suppose that $g(x, y)$ encodes the probability distribution over labels $\mathcal{Y}$ given a feature vector x, i.e., $g(x, y) = \Pr\{y|x\}$ (this is a common case for deep neural network models for image classification, for example). Then $f(x)$ becomes the most probable label $y \in \mathcal{Y}$.

2.1.3   PAC LEARNABILITY

Probably approximately correct (PAC) learnability is an important theoretical framework for (most supervised) machine learning. Here, we present the idea specifically for binary classification. Formally, let $\mathcal{F}$ be the class of possible classifiers which the defender considers (i.e., the defender's *hypothesis* or *model class*). Let $(x, y) \sim \mathcal{P}$ be instances, where $x \in X$ is an input feature vector, and y is a label in $\{0, 1\}$. To simplify exposition, suppose that $y = h(x)$ for some function $h(x)$ not necessarily in $\mathcal{F}$ (i.e., output is a deterministic function of input x, for example, the true classification of x as either benign or malicious). For any $f \in \mathcal{F}$, let $e(f) = \Pr_{x \sim \mathcal{P}}[f(x) \neq h(x)]$ be the expected error of f w.r.t. $\mathcal{P}$, and we define $e_{\mathcal{F}} = \inf_{f \in \mathcal{F}} e(f)$ as the optimal (smallest) error achievable by any function $f \in \mathcal{F}$. Let $z^m = \{(x_1, y_1), \ldots, (x_m, y_m)\}$ be data generated according to $\mathcal{P}$ and let Z^m be the set of all possible z^m.

Definition 2.1   Let $\mathcal{F}$ be a class of functions mapping x to $\{0, 1\}$. A *learning algorithm* is a function $L : \cup_{m \geq 1} Z^m \to \mathcal{F}$. We say that $\mathcal{F}$ is *PAC learnable* if there is a learning algorithm for

$\mathcal{F}$ with the property that for any $\epsilon, \delta \in (0, 1)$ and any $\mathcal{P}$, there exists $m_0(\epsilon, \delta)$, such that for all $m \geq m_0(\epsilon, \delta)$, $\mathrm{Pr}_{z^m \sim \mathcal{P}}\{e(L(z^m)) \leq e_{\mathcal{F}} + \epsilon\} \geq 1 - \delta$. We say it is efficiently (polynomially) PAC learnable if $m_0(\epsilon, \delta)$ is polynomial in $1/\epsilon$ and $1/\delta$ and there exists a learning algorithm for $\mathcal{F}$ which runs in time polynomial in m, $1/\epsilon$, and $1/\delta$.[1] We say that $m_0(\epsilon, \delta)$ is this algorithm's *sample complexity*. Henceforth, we will often omit the *PAC* modifier, and just use the term *learnable* to mean *PAC learnable*. We will say that an algorithm that can obtain the PAC guarantees is a *PAC learning algorithm*; if the algorithm runs in polynomial time, and has polynomial sample complexity, we call it a *polynomial PAC learning algorithm*.

2.1.4 SUPERVISED LEARNING IN ADVERSARIAL SETTINGS

Regression Learning in Adversarial Settings   An example of the use of regression learning in adversarial settings would be a parametric controller learned from observations of actual control decisions—as has been shown in the context of autonomous driving. To simplify, suppose that we learn a controller $f(x)$ to predict a steering angle as a function of vision-based input (Figure 2.2, left) captured into a feature vector x, as is done in end-to-end autonomous driving [Bojarski et al., 2016, Chen and Huang, 2017]. The adversary may introduce small manipulations into the image captured by the vision system, thereby modifying x to x' to introduce an error in the predicted steering angle $f(x')$ to maximize the difference from the true optimal angle y (Figure 2.2, right).

Attack
$\Longrightarrow$

Figure 2.2: Screenshots of Udacity car simulations. Left: image input into the autonomous controller. Right: car veers off the road after an attack which compromises the operation of the controller.

As another example, the learner may wish to predict the stock price $f(x)$ as a function of observables x. An adversary, aspiring to profit from the mistakes by the learner, may attempt to influence the observed state x which is used to predict stock price by manipulating it into another, x', so that the predicted price in the next period is high. This may result in the learner

---

[1]This definition is taken, in a slightly extended form, from Anthony and Bartlett [1999, Definition 2.1].

willing to buy the stock from the adversary at an inflated price, resulting in an effective arbitrage opportunity for the adversary at the expense of the learner.

Classification Learning in Adversarial Settings   Applying binary classification in adversarial settings commonly amounts to distinguishing between benign and malicious instances. In email filtering, for example, malicious instances would be spam, or phishing emails, while benign would be regular email traffic [Bhowmick and Hazarika, 2018]. In malware detection, malicious entities would be, naturally, malicious software, while benign instances would correspond to non-malicious executables [Chau et al., 2011, Smutz and Stavrou, 2012, Šrndić and Laskov, 2016, Tamersoy et al., 2014, Ye et al., 2017]. In credit card fraud, one would consider specific features of credit card applications to determine whether the application is fraudulent (malicious) or legitimate (benign) [Lebichot et al., 2016, Melo-Acosta et al., 2017]. In all these cases, the adversary has an incentive to avoid being detected, and would wish to manipulate their behavior so that it appears benign to the detector.

2.2   UNSUPERVISED LEARNING

In unsupervised learning a dataset is comprised of only the feature vectors, but has no labels: $\mathcal{D} = \{x_i\}$. Consequently, problems in unsupervised learning are concerned with identifying aspects of the joint distribution of observed features, rather than predicting a target label. However, the line between supervised and unsupervised techniques may at times blur, as is the case with matrix completion methods, where the goal is to predict unobserved matrix entries.

There are a number of specific problems that are commonly studied under the unsupervised learning umbrella. We discuss three of these: *clustering*, *principal component analysis*, and *matrix completion*.

2.2.1   CLUSTERING

One of the most familiar examples of unsupervised learning is the *clustering* task, in which the feature vectors in a dataset are divided into a collection of subsets $\mathcal{S}$, such that feature vectors in each collection $S \in \mathcal{S}$ are "close" to the mean feature vector of S for some measure of closeness.

Formally, clustering can be seen as solving the following optimization problem:

$$\min_{\mathcal{S},\mu} \sum_{S \in \mathcal{S}} \sum_{i \in S} l(x_i, \mu_S),$$

where $\mathcal{S}$ is a partition of $\mathcal{D}$, and μ_S an aggregation measure of the data in cluster $S \in \mathcal{S}$, for example, its mean. A common version of this problem uses the l_2 norm as the loss function, and a heuristic approximation of the associated problem is k-means clustering, where one iteratively updates cluster means and moves data points to a cluster with the closest mean. However, other variations are possible, and regularization can also be added to this problem to control model complexity (such as the number of clusters formed, if this number is not specified up front).

A more general approach is to learn a distribution over the dataset $\mathcal{D}$. A well-known example is a Gaussian Mixture Model, in which x_i are assumed to be sampled i.i.d. from a density which is a linear mixture of multi-variate Gaussian distributions. This approach is also known as "soft clustering," since each Gaussian in the mixture can be assumed to generate data in a particular "cluster," but we allow for uncertainty about cluster membership.

2.2.2   PRINCIPAL COMPONENT ANALYSIS

Principal component analysis (PCA) finds a collection of $K < m$ orthonormal basis vectors $\{v_k\}_{k=1}^{K}$ which are the k eigenvectors of the data matrix $\mathbf{X}$, where each feature vector x_i in the dataset $\mathcal{D}$ is a row in this matrix. Equivalently, each v_k solves

$$v_k = \arg\max_{v:\|v\|=1} \|\mathbf{X}(\mathbb{I} - \sum_{i=1}^{k-1} v_i v_i^T)v\|,$$

where $\mathbb{I}$ is the identity matrix.

Let $\mathbf{V}$ be the basis matrix produced by PCA, in which columns correspond to the eigenvectors v_k. Then for any feature vector x, its m-dimensional reconstruction is

$$\tilde{x} = \mathbf{V}\mathbf{V}^T x$$

and the corresponding residual error (i.e., the error resulting from using PCA to approximate the original feature vector x) is

$$x_e = x - \tilde{x} = (\mathbb{I} - \mathbf{V}\mathbf{V}^T)x.$$

Intuitively, we expect PCA to be effective if the original data can be effectively represented in a K-dimensional subspace (where K is small relative to the original dimension m); in other words, the magnitude of the residual error $\|x_e\|$ is small.

2.2.3   MATRIX COMPLETION

To motivate the problem of *matrix completion*, consider a collaborative filtering problem faced by Netflix. The goal is to predict how much a given user would like (or, more accurately, rate) a given movie. One way to represent this problem is to imagine a large matrix $\mathbf{M}$ with n rows corresponding to users and m columns to movies, where each matrix entry is the rating a user i would give to a movie j. Thus, our prediction problem can be equivalently viewed as predicting the value of the ijth entry of this large matrix, which is otherwise very sparsely populated from actual movie ratings provided by the users. The key insight is to hypothesize that the true underlying matrix of ratings is low-rank; for example, movies tend to be rated similarly by similar users. Consequently, we can decompose the true matrix as $\mathbf{M} = \mathbf{U}\mathbf{V}^T$, where $\mathbf{U}$ and $\mathbf{V}$ have K columns, and K is the rank of $\mathbf{M}$. Our goal would be to obtain $\mathbf{U}$ and $\mathbf{V}$ such that $U_i V_j^T$ are good approximations of observed entries of $\mathbf{M}$, M_{ij} (where U_i is the row vector denoting the ith row of $\mathbf{U}$ and V_j the row vector corresponding to the jth row of $\mathbf{V}$).

To formalize, let $\mathbf{M} \in \mathbb{R}^{n \times m}$ be a data matrix consisting of m rows and n columns. M_{ij} for $i \in [n]$ and $j \in [m]$ would then correspond to the rating the ith user gives for the jth item. We use $\Omega = \{(i, j) : M_{ij}$ is observed$\}$ to denote all observed entries in $\mathbf{M}$ and assume that $|\Omega| \ll mn$. We also use $\Omega_i \subseteq [m]$ and $\Omega_j' \subseteq [n]$ for columns (rows) that are observed at the ith row (jth column). The goal of *matrix completion* is to recover the complete matrix $\mathbf{M}$ from few observations $\mathbf{M}_\Omega$ [Candès and Recht, 2007].

The matrix completion problem is in general ill-posed as it is impossible to complete an arbitrary matrix with partial observations. As a result, additional assumptions are imposed on the underlying data matrix $\mathbf{M}$. One standard assumption is that $\mathbf{M}$ is very close to an $n \times m$ rank-K matrix with $K \ll \min(n, m)$. Under such assumptions, the complete matrix $\mathbf{M}$ can be recovered by solving the following optimization problem:

$$\min_{\mathbf{X} \in \mathbb{R}^{n \times m}} \|\mathcal{R}_\Omega(\mathbf{M} - \mathbf{X})\|_F^2, \quad s.t. \ \text{rank}(\mathbf{X}) \leq K, \tag{2.5}$$

where $\|\mathbf{A}\|_F^2 = \sum_{i,j} \mathbf{A}_{ij}^2$ denotes the squared Frobenious norm of matrix $\mathbf{A}$ and $[\mathcal{R}_\Omega(\mathbf{A})]_{ij}$ equals $\mathbf{A}_{ij}$ if $(i, j) \in \Omega$ and 0 otherwise. Unfortunately, the feasible set in Eq. (2.5) is non-convex, making the optimization problem difficult to solve. There has been an extensive literature on approximately solving Eq. (2.5) and/or its surrogates that lead to two standard approaches: alternating minimization and nuclear norm minimization. For the first approach, one considers the following problem:

$$\min_{\mathbf{U} \in \mathbb{R}^{n \times K}, \mathbf{V} \in \mathbb{R}^{m \times K}} \left\{ \|\mathcal{R}_\Omega(\mathbf{M} - \mathbf{U}\mathbf{V}^T)\|_F^2 + 2\gamma_U \|\mathbf{U}\|_F^2 + 2\gamma_V \|\mathbf{V}\|_F^2 \right\}. \tag{2.6}$$

Equation (2.6) is equivalent to Eq. (2.5) when $\gamma_U = \gamma_V = 0$. In practice, people usually set both regularization parameters γ_U and γ_V to be small positive constants in order to avoid large entries in the completed matrix and also improve convergence. Since Eq. (2.6) is bi-convex in $\mathbf{U}$ and $\mathbf{V}$, an *alternating minimization* procedure can be applied, in which we iteratively optimize with respect to $\mathbf{U}$ and $\mathbf{V}$ (each optimization problem is convex if we fix the other matrix).

Alternatively, one solves a *nuclear-norm minimization* problem

$$\min_{\mathbf{X} \in \mathbb{R}^{n \times m}} \|\mathcal{R}_\Omega(\mathbf{M} - \mathbf{X})\|_F^2 + 2\gamma \|\mathbf{X}\|_*, \tag{2.7}$$

where $\gamma > 0$ is a regularization parameter and $\|\mathbf{X}\|_* = \sum_{i=1}^{\text{rank}(\mathbf{X})} |\sigma_i(\mathbf{X})|$ is the nuclear norm of $\mathbf{X}$, which acts as a convex surrogate of the rank function. Equation (2.7) is a convex optimization problem and can be solved using an iterative singular value thresholding algorithm [Cai et al., 2010]. It can be shown that both methods in Eq. (2.6) and (2.7) provably approximate the true underlying data matrix $\mathbf{M}$ under certain conditions [Candès and Recht, 2007, Jain et al., 2013].

2.2.4   UNSUPERVISED LEARNING IN ADVERSARIAL SETTINGS

Two common uses of unsupervised learning in adversarial settings are attack clustering and anomaly detection.

Clustering   Clustering of attacks can be useful in malware forensic analysis, where different malware variants are clustered, for example, to determine whether they are from the same family [Hanna et al., 2013, Perdisci et al., 2013]. The technique could be important in identifying variants of the same malware in cases where malware polymorphism had been used to hide the malware from anti-virus tools. It can also be valuable in determining the origins of the malware (such as malware authors).

Anomaly Detection   In anomaly detection, one typically uses a collection of "normal" operational data to develop a model of "normal" system behavior. For example, one can collect traces of routine network traffic in an organization, and identify a collection of statistics (features) of this "normal" traffic. The ultimate goal is to identify behavior which is anomalous, suggesting that it is either due to a system fault (in non-adversarial settings) or an attack, such as an intrusion. A number of variations of anomaly detection exist. Here we describe several concrete examples which have been used in adversarial settings, and have also been subject to attacks.

One simple and surprisingly general anomaly detection approach is *centroid anomaly detection* [Kloft and Laskov, 2012]. In this approach, one uses training data $\mathcal{D}$ to obtain a mean, $\mu = \frac{1}{n} \sum_{i \in \mathcal{D}} x_i$, and labels any new feature vector x, as anomalous if

$$\|x - \mu\|_p^p \geq r,$$

where r is an exogenously specified threshold, typically set to limit a false positive rate below a target level. If $p = 2$ (as is common), we can rewrite the difference between x and μ as

$$
\begin{aligned}
\|x - \mu\|_2^2 &= \langle x - \mu, x - \mu \rangle = \langle x, x \rangle - 2\langle x, \mu \rangle + \langle \mu, \mu \rangle \\
&= \langle x, x \rangle - \frac{2}{n}\langle x, \sum_i x_i \rangle + \langle \frac{1}{n}\sum_i x_i, \frac{1}{n}\sum_i x_i \rangle \\
&= \langle x, x \rangle - \frac{2}{n}\sum_i \langle x, x_i \rangle + \frac{1}{n^2}\sum_{i,j} \langle x_i, x_j \rangle \\
&= k(x, x) - \frac{2}{n}\sum_i k(x, x_i) + \frac{1}{n^2}\sum_{i,j} k(x_i, x_j),
\end{aligned}
$$

where $k(\cdot, \cdot)$ is a kernel function which can represent a dot-product in a higher-dimensional space and $\langle x, y \rangle$ represents the dot product of vectors x and y, and allows us to consider complex nonlinear extensions of the original simple centroid approach.

Another useful property of a centroid anomaly detector is that we can update the mean μ online as new (normal) data arrives. Formally, suppose that μ_t is the mean computed from past data, and a new data point x_t has just arrived. The new estimate μ_{t+1} can then be computed as

$$\mu_{t+1} = \mu_t + \beta_t(x_t - \mu_t),$$

where β_t is the learning rate.

Another important class of anomaly detectors leverage an observation that normal behavior (such as network traffic) often has a low intrinsic dimension [Lakhina et al., 2004]. The main idea is to use PCA to identify a high-quality low-dimensional representation of the data, and use the magnitude of the residual to determine anomalies. Such an anomaly detector flags an observation x if the norm of its residual x_e is too large (above a predefined threshold):

$$\|x_e\| = \|(\mathbb{I} - \mathbf{V}\mathbf{V}^T)x\| \geq r.$$

An example use case of PCA-based anomaly detectors is to identify anomalous traffic flow activity [Lakhina et al., 2004]. The setting they consider is where PCA-based anomalies correspond to unusual origin-destination (OD) network flows. Specifically, let a matrix $\mathbf{A}$ represent which OD flow uses which links; that is, $A_{if} = 1$ if flow f uses link i and 0 otherwise. Let a matrix $\mathbf{X}$ represent observed flows over time, so that X_{tf} is the amount of traffic over an OD flow f at time period t. Then we can define $\mathbf{Y} = \mathbf{X}\mathbf{A}^T$ to represent time dynamics of flow over individual links, with $y(t)$ the corresponding flow over links at time period t. If we assume that $\mathbf{Y}$ is approximately K-rank, PCA would output the top K eigenvectors of $\mathbf{Y}$. Let $\mathbf{V}$ represent the associated matrix computed using PCA, as above. New flows y' can then be determined as anomalous or not by computing the residual $y_e = \|(\mathbb{I} - \mathbf{V}\mathbf{V}^T)y\|$, and comparing it to the threshold r. Lakhina et al. [2004] use the Q-statistic to determine the threshold at a predefined $1 - \beta$ confidence level.

A third variant of anomaly detection techniques uses n-grams, or sequences of n successive entities in an object, to determine whether it is similar to a statistical model of normal objects. To be more precise, suppose that we use the n-grams approach for anomaly-based network intrusion detection. We would start by collecting network data, and developing a statistical model of normal. Let's suppose that our analysis is at the level of packets (in practice, one can also analyze sessions, files, etc). Ideally, we would use a feature representation with features corresponding to all possible n-grams, and feature values corresponding to the frequency with which the associated n-gram had been observed. For a new packet, we can then use a sliding window to obtain its n-gram frequencies, and use conventional centroid anomaly detection to trigger alerts. As this idea is not scalable when n becomes large, Wang et al. [2006] suggest instead storing n-grams observed during training using a Bloom filter, and scoring only based on the proportion of *new* n-grams contained in the packet at test time.

2.3   REINFORCEMENT LEARNING

We start by describing Markov Decision Processes (MDPs), which provide the mathematical foundations for reinforcement learning (RL) [Sutton and Barto, 1998].

A *discrete-time discounted infinite-horizon* MDP (which we focus on for simplicity of the exposition) is described by a tuple $[S, A, T, r, \delta]$ where S is a set of states (which we assume is finite here), A the set of actions (again, assumed finite), T the transition dynamics, where $T_{ss'}^a = \Pr\{s_{t+1} = s' | s_t = s, a_t = a\}$, $r(s, a)$ the expected reward function, and $\delta \in [0, 1)$ the discount factor. Two central concepts in MDPs and RL are the value function, defined as

$$V(s) = \max_a \left(r(s, a) + \delta \sum_{s'} T_{ss'}^a V(s') \right)$$

and the Q-function, defined as

$$Q(s, a) = r(s, a) + \delta \sum_{s'} T_{ss'}^a V(s').$$

Conceptually, the value function captures the optimal discounted sum of rewards one can obtain (i.e., if we follow an optimal policy), while the Q-function is the discounted reward if one takes an action a in state s and follows an optimal policy thereafter. Notice that $V(s) = \max_a Q(s, a)$. There are a series of well-known ways to compute an optimal policy for MDPs, where a policy is a mapping $\pi : S \rightarrow A$ from states to actions.[2] One example is value iteration, in which one transforms the characterization of the value function above into an iterative procedure computing value function $V_{i+1}(s)$ in iteration $i + 1$ as

$$V_{i+1}(s) = \max_a \left(r(s, a) + \delta \sum_{s'} T_{ss'}^a V_i(s') \right).$$

We can initialize this process with an arbitrary value function, and it will always converge to the true value function in the limit. Another alternative is to use policy iteration, where one alternates policy evaluation (computing the value of the current policy) and improvement steps (analogous to value iteration steps). Finally, one can compute the optimal value function using linear programming. One can extract the optimal policy from a value function by simply finding the maximizing action in each state.

In reinforcement learning, one knows S, A, and δ, but not T or r. However, we can nevertheless learn from experience to obtain, eventually, a policy which is close to optimal. A number of algorithms exist for this purpose, the best known of which is perhaps Q-learning. In Q-learning, one would initialize the Q-function in an arbitrary way. In an arbitrary iteration $i + 1$, one has observed a state s_i and takes an action a_i according to some current policy which ensures that any action can be taken in any state with positive probability. If we then observe a reward r_{i+1} and next state s_{i+1}, we can update the Q-function as follows:

$$Q_{i+1}(s_i, a_i) = Q_i(s_i, a_i) + \beta_{i+1}(r_{i+1} + \delta \max_a Q_i(s_{i+1}, a) - Q_i(s_i, a_i)).$$

[2]For infinite-horizon discrete-time discounted MDPs, there always exists an optimal policy that is only a function of observed state.

Now, a naive way to compute a policy based on this Q-function would be to simply take the action in each state which maximizes $Q_i(s, a)$ in the current iteration i. However, this would mean that no exploration takes place. One simple modification, called ϵ_i-greedy, would be to use such a policy with probability $1 - \epsilon_i$, and play a random action otherwise. ϵ_i could then be decreased over time. Another idea would be to use the policy which plays an action a with probability

$$\pi_i(s, a) \propto \beta_i Q_i(s, a),$$

with β_i increasing over time.

When RL is applied in structured domains (characteristic of many real applications), it is common for states to be comprised of a collection of variables $x = \{x_1, \ldots, x_n\}$ (this is often known as a *factored representation of state*). In this case, the policy cannot be reasonably represented as a lookup table. There are several common approaches to handle this. For example, one can learn a parametric representation of the Q-function with parameters w, $Q(x, a; w)$, and use it to indirectly represent the policy, where $\pi(x) = \arg\max_a Q(x, a; w)$.

2.3.1   REINFORCEMENT LEARNING IN ADVERSARIAL SETTINGS

One example of the use of reinforcement learning in (potentially) adversarial settings would be autonomous control (e.g., self-driving cars) in the presence of adversaries who may impact observed state. Since the optimal policy depends on state, modifications to state may then result in poor decisions. Specific formal models of such attacks would leverage the factored state representation: the attacker would typically only be able to modify a (small) subset of state variables to lead a learned policy astray.

Indeed, such attacks actually are not specific to RL, and simply attack a particular policy used (if known to the attacker). One could also attack during the learning process, to cause a poor policy to be learned—for example, by affecting the observed rewards, as well as observed states. In any case, the upshot of the attack would be to cause the RL-based autonomous agent to make mistakes, potentially in a high-stakes situation, such as causing an autonomous car to crash.

2.4   BIBLIOGRAPHIC NOTES

Our discussion of machine learning generally, as well as that of specific techniques, is informed by a number of well-known books on machine learning in general [Bishop, 2011, Hastie et al., 2016], as well as the unparalleled text on reinforcement learning by Sutton and Barto [1998]. A foundational perspective on machine learning, including PAC learnability, can be found in a wonderful theoretical treatment by Anthony and Bartlett [2009]. Similarly, Vapnik [1999] offers a great foundational discussion of empirical risk minimization.

Our description of centroid-based anomaly detection is based on Kloft and Laskov [2012], while we follow Lakhina et al. [2004] in discussing PCA-based anomaly detection. Our discussion of matrix completion closely follows Li et al. [2016], who in turn build on extensive prior research on the topics of matrix factorization and completion [Candès and Recht, 2007, Gemulla et al., 2011, Gentle, 2007, Sra and Dhillon, 2006].

A classic text on function approximation in markov-decision processes and reinforcement learning is by Bertsekas and Tsitsiklis [1996], with numerous advances improving and extending both the techniques and theoretical foundations [Boutilier et al., 1999, 2000, Guestrin et al., 2003, St-Aubin et al., 2000].

CHAPTER 3

Categories of Attacks on Machine Learning

In the previous chapter, we described in broad strokes the major machine learning paradigms, as well as how they can be instantiated in adversarial settings. Adversarial machine learning takes this a step further: our goal is not merely to understand how machine learning can be used in adversarial settings (for example, for malware detection), but in what way such settings *introduce vulnerabilities* into conventional learning approaches. A principled discussion of such vulnerabilities centers around precise threat models. In this chapter, we present a general categorization of threat models, or attacks, in the context of machine learning. Our subsequent detailed presentation of the specific attacks will be grounded in this categorization.

There have been several attempts at categorizing attacks on machine learning algorithms. The categorization we propose is related to some of these, and aims to distill the most important features of the attacks that we discuss. In particular, we classify attacks along three dimensions: *timing*, *information*, and *goals*.

1. **Timing:** The first crucial consideration in modeling attacks is *when* the attack takes place. This consideration leads to the following common dichotomy which is central to attacks on machine learning: attacks on models (of which *evasion attacks* are the most prototypical cases) and attacks on algorithms (commonly known as *poisoning* attacks). Attacks on models or, more precisely on *decisions* made by learned models, assume that the model has already been learned, and the attacker now either changes its behavior, or makes changes to the observed environment, to cause the model to make erroneous predictions. Poisoning attacks, in contrast, take place before models are trained, modifying a part of the data used for training. This distinction is illustrated in Figure 3.1.

2. **Information:** The second important issue in modeling attacks is what information the attacker has about the learning model or algorithm, a distinction which is commonly distilled into *white-box* vs. *black-box* attacks. In particular, white-box attacks assume that either the model (in the case of attacks on decisions) or algorithm (in poisoning attacks) is fully known to the adversary, whereas in black-box attacks the adversary has limited or no information about these, although may obtain some of the information indirectly, for example, through queries.

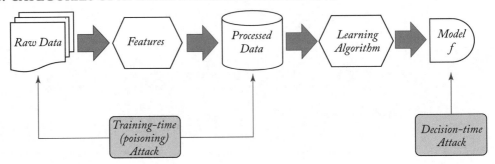

Figure 3.1: A schematic representation of the distinction between decision-time attacks (attacks on models) and poisoning attacks (attacks on algorithms).

3. **Goals:** Attackers may have different reasons for attacking, such as evading detection or reducing confidence in the algorithm. We differentiate two broad classes of attack goals: *targeted attacks* and *attacks on reliability of the learning method* (or simply *reliability attacks*). In a targeted attack, the attacker's goal is to cause a mistake on specific instances of a specific nature (for example, causing a learned function f to predict a specific erroneous label l on an instance x). A reliability attack, in contrast, aims to degrate the perceived reliability of the learning system by maximizing prediction error.

In Table 3.1 we summarize our categorization of attacks on machine learning.

Table 3.1: The three dimensions of attacks on machine learning

Attack Timing	Decision time (e.g., evasion attack) vs. training time (poisoning)
Attacker Information	White-box vs. black-box attacks
Attack Goals	Targeted attacks vs. reliability attacks

In the remainder of this chapter we discuss these three dimensions of attacks in greater depth. The high-level organization of the book, however, centers largely on the first dichotomy between attacks on models and poisoning attacks, which we view as the most fundamental distinction.

3.1   ATTACK TIMING

Attacks at Decision Time: Of all the attack classes we will consider, *evasion* attacks—a major subclass of attacks which take place at decision time—are perhaps the most salient historically. A well-known example of evasion is the evolution of spam email traffic, for example, when spammers replace the letter "i" with a number "1" or a letter "l" in "Viagra" (which becomes "V1agra").

In general, a *classifier evasion* attack on binary classifiers takes as input a classifier $f(x)$ and an "ideal" instance in feature space, x_{ideal}, (i.e., this is what the adversary would wish to do if there were no classifier to identify it as malicious). The attack then outputs another instance, corresponding to a feature vector x'. If $f(x') = -1$, evasion is successful, but it is possible that the adversary fails to find an adequate evasion (in fact, it is necessary for any meaningful measure of evasion-robustness of an algorithm that the adversary does not *always* succeed to find an evasion, no matter what f and x_{ideal} are).

As an illustration, let us say that one wishes to detect spam email, and learns a classifier $f(x)$ for this purpose (where x is a vector representing features of an email). Now consider a spammer who previously used a template corresponding to a feature vector x_{spam} and suppose that $f(x_{spam})$ labels it as "spam" $(+1)$, so that the spammer receives no responses. The spammer would make modifications to the email to obtain an instance which in feature space looks like x' with the property that $f(x') = -1$ (i.e., it's classified as non-spam, and allowed to pass into users' mailboxes). But x' cannot be arbitrary: the adversary incurs a cost of modifying the original instance x_{spam} to achieve x', which could measure the cost of effort (to maintain functionality), or effectiveness (for example, they may have to introduce spelling errors which would allow the attacker to avoid detection, but would also reduce the chance that people will click on embedded links).

Generalizing the idea behind evasion attacks, we can consider decision-time attacks on multi-class classification. Let $\mathcal{Y}$ be a finite set of labels, and suppose that for some instance x_{ideal} the predicted label is $f(x_{ideal}) = y$. An attacker may wish to change this instance into another, x', either to effect an incorrect prediction $(f(x') \neq y)$, or to cause the classifier to predict a target label $t = f(x')$. Recently, such attacks have gained a great deal of attention under the term *adversarial examples*, largely focusing on vision applications and deep neural networks. A potential concern is that an attacker may attempt to cause an autonomous vehicle relying on vision to crash by manipulating the perceived image of a road sign, such as a stop sign (for example, by posting specially crafted stickers that appear to be graffiti to passers by [Evtimov et al., 2018]). While this problem is a special case of decision-time attacks, the amount of independent attention it has received warrants a separate chapter (Chapter 8).

Attacks on Training Data: The issue of learning with corrupted or noisy training data has been a subject of considerable investigation in the machine learning and statistics communities for several decades. However, *adversarial* corruption of training data has received more systematic consideration recently, particularly if we allow a non-negligible proportion of data to be corrupted. The nature of poisoning attacks is that the adversary deliberately manipulates the *training data* prior to training to cause the learning algorithm to make poor choices. An important conceptual challenge with poisoning attacks is in defining the scope of adversarial manipulation of training data, and the goals of the adversary in doing so. One common way to sidestep these issues is to assume that the adversary can make arbitrary modifications to a small subset of training data points. The goal would then be to design algorithms which are robust to such arbitrary training data corruption, as long as the amount of corrupted data is sufficiently small.

One can also consider more specific models of corruption which impose additional constraints on what the attacker may do. One common class of such attacks are *label-flipping* attacks, where the adversary is allowed to change the labels of at most C data points in the training data. Typically, such attacks have been considered in the context of classification, although one can also modify regression labels. In most cases, one assumes that the algorithm and feature space are known to the adversary (i.e., a *white-box attack*). Data poisoning can also be considered in unsupervised learning settings, for example, when it is used in anomaly detection. In this case, the adversary may make small modifications to observed normal behavior which now pollutes the model used to detect anomalies, with the goal of ensuring that a future target attack is flagged as benign.

3.2   INFORMATION AVAILABLE TO THE ATTACKER

One of the most important factors in attack modeling is information that the attacker has about the system they are attacking. We draw a distinction between *white-box attacks*, in which the attacker knows everything there is to know, and *black-box attacks*, in which the attacker has limited information.

White-box attacks assume that the adversary knows *exactly* either the learned model (e.g., actual classifier) in the case of decision-time attacks, or the learning algorithm in the case of poisoning attacks. This means, for example, that the adversary knows all the model parameters, including features, and, in the case of poisoning attacks, the hyperparameters of the learning algorithm. The assumption that the attacker has such intimate information about the learning system can seem suspect. There are, however, important reasons to consider white-box attacks. First, these offer a natural starting point from a learner's perspective: if a learner can be robust to white-box attacks, they are surely robust also to attacks which are informationally limited. Second, from an attacker's perspective, there may be a number of ways to indirectly obtain sufficient information about a learned model to deploy a successful attack. Take a malware evasion attack for example. Suppose that the set of features used is public information (such as through published work), and datasets used to train a malware detector are public (or, alternatively, there are public datasets which are sufficiently similar to the data actually used for training). Finally, suppose that the learner uses a standard learning algorithm to learn the model, such as a random forest, deep neural network, or support vector machine, and standard techniques to tune hyperparameters, such as cross-validation. In this case, the attacker can obtain an identical, or nearly identical, version of the detector as the one in actual use!

In **black-box attacks**, in contrast to white-box attacks, the adversary does not have precise information about either the model or the algorithm used by the learner. An important modeling challenge for black-box attacks is to model precisely *what* information the attacker has about either the learned model, or the algorithm.

In the context of decision-time black-box attacks, one approach is to consider a hierarchy of information about the learned model available to the adversary. At one extreme, no information is available to the adversary at all. A more informed adversary may have some training data which is different from the data on which the actual model had been trained, but no information about the particular model class being learned, or features used. A more informed attacker yet may know the model class and features, and perhaps the learning algorithm, but have no training data, and an even more informed adversary may also have training data sampled from the same distribution as the data used for learning. Finally, when this same adversary has the *actual* training data used by the learning algorithm, the resulting attack is equivalent to a white-box attack as discussed above, since the attacker can learn the exact model from the given training data. One may observe that, unlike white-box attacks, there are many ways to model black-box attacks. Indeed, others have suggested a term *gray-box attack* to indicate that the attacker has some, albeit incomplete, information about the system they are attacking [Biggio and Roli, 2018]. In this book, however, we will keep with the more conventional term of black-box attacks to refer to the entire hierarchy, short of the white-box full information extreme.

The information hierarchy above does not address a natural question: how does the attacker come by the information about the model they are attacking (in the case of decision-time attacks)? An important class of decision-time black-box attack models addresses this question by allowing an attacker to have query access to the learned model. Specifically, given an arbitrary instance, represented as a feature vector x, the adversary can obtain (query) the actual predicted label $y = f(x)$ for the unknown black-box model f. Commonly, and implicitly, such query models also assume that the attacker knows both the model space (e.g., learning algorithm) and feature space. Moreover, another practical limitation of this model is that $f(x)$ is often imprecisely, or noisily observed given x. For example, suppose that the spammer sends a spam email message. A non-response need not imply that it has been filtered—rather, it could be that users simply disregarded the email. Nevertheless, such a query-based framework enables an elegant theoretical study of what an attacker can accomplish with very limited information about the learner.

Following similar principles as typical models of black-box attacks at decision-time, black-box data poisoning attacks would allow for a range of knowledge about the algorithm used by the defender. For example, at one extreme, the attacker may have no information at all about the algorithm. A more informed attacker may know the algorithm, but not the hyperparameters (such as the regularization weight, or the number of hidden layers in the neural network) or features. A more informed attacker yet may know the algorithm, features, and hyperparameters, but not the training data that the attacker attempts to poison.

3.3 ATTACKER GOALS

While attackers may have a broad number of possible goals for perpetrating an attack on the machine learning systems, we distill attacks into two major categories in terms of attacker goals: *targeted attacks* and *reliability attacks*.

Targeted attacks are characterized by a specific goal for the attacker with regard to model decisions. For example, consider a decision-time attack on a multi-class classifier with a set of possible labels $\mathcal{L}$, and let x be a particular instance of interest to an attacker with a true label y. A goal of a targeted attack in this case would be to effect a change in the label for x to a *specific* target label $t \neq y$. More generally, a targeted attack is characterized by a subset of the joint instance and (if relevant) label space $S \subseteq (\mathcal{X} \times \mathcal{Y})$ of datapoints that the attacker would wish to change the decision for, along with target decision function $D(x)$. In the most common setting of targeted attacks on supervised learning, an attacker would aspire to induce predictions on each $(x, y) \in S$ to be according to a target label function $l(x)$.

Reliability attacks, on the other hand, attempt to maximize mistakes in decisions made by learning with respect to ground truth. For example, in supervised learning, an attacker would aim to maximize prediction error. In vision applications, such attacks, which have come to be commonly called *untargeted*, would modify an image so as to cause an erroneous prediction (e.g., recognition of an object not in the image, such as mistaking an image of a stop sign for any other road sign).

Our distinction between targeted and reliability attacks blurs when we consider binary classification: in particular, reliability attacks now become a special case in which the target labels $l(x)$ are simply the alternative labels. More generally, we note that even the dichotomy between targeted and reliability attacks is incomplete: for example, one may consider attacks in which the goal is to avoid predictions of a particular class *other* than the correct label (sometimes called a *repulsive attacks*). However, this problem is sufficiently esoteric that we feel justified in focusing on our simplified categorization.

3.4 BIBLIOGRAPHIC NOTES

Barreno et al. [2006] presented the first taxonomy of attacks on machine learning, while Barreno et al. [2010] elaborate on this taxonomy to present a comprehensive categorization of such attacks. This taxonomy also considers three dimensions of attacks. Their first dimension is essentially identical to ours, albeit they identify it with attacker *influence*, rather than attack timing. The associated dichotomy in this category is *causative* vs. *exploratory* attacks. Causative attacks described by Barreno et al. [2010] are identical to what we call *poisoning* attacks, whereas their *exploratory* attacks seem largely aligned with our *decision-time* attacks. Their second and third dimensions, however, are somewhat different from our categorization. The second dimension in Barreno et al. [2010] is termed *security violation*, and distinguishes *integrity* and *availability* attacks. Integrity attacks are those which cause false negatives (that is, which cause malicious instances to remain undetected), whereas availability attacks cause denial of service through false positives (i.e., benign instances which are flagged as malicious as a result of the attack). We can note that this category appears to be quite specific to binary classification, with an eye largely to detection of malicious instances such as spam. The final dimension identified by Barreno et al. [2010] is *specificity*, and here they differentiate between *targeted* and *indiscriminate* attacks. The notion of targeted attacks is similar to ours, while indiscriminate attacks are similar to what we call *reliability* attack. The key difference is that indiscriminate attacks in Barreno et al. [2010] terminology are broader: for example, these allow for a spammer to target a large set of *spam* instances (but not benign instances), whereas the attacker's goal in reliability attacks (according to our definition) is simply to maximize prediction error on *all* instances. Thus, we essentially collapse two of the dimensions from Barreno et al. [2010] into the simpler dichotomy of two attack goals (targeted vs. reliability attacks), and add a new dimension which distinguishes white-box and black-box attacks.

Our categorization of black-box attacks at decision time is informed by the query-based models of classifier evasion attacks [Lowd and Meek, 2005a, Nelson et al., 2012], as well as the attacker information hierarchy elaborated by Biggio et al. [2013] and Šrndic and Laskov [2014]. An analogous characterization of black-box poisoning attacks (somewhat applicable also to attacks at decision time) is due to Suciu et al. [2018], who call it *FAIL* after their four dimensions of attacker knowledge: **F**eature knowledge (which features are known to the attacker), **A**lgorithm knowledge (to what extent does the attacker know the learning algorithm), **I**nstance knowledge (what information the attacker has about the learner's training data), and **L**everage (which features the attacker can modify). Finally, our categorization is also informed by a recent survey on adversarial machine learning by Biggio and Roli [2018], and more generally by the broader literature on adversarial machine learning, including attacks specifically on deep learning models. We present bibliographic notes on these efforts in later chapters, where we discuss attacks on machine learning, and associated defenses, in greater depth.

CHAPTER 4

Attacks at Decision Time

In this chapter, we begin to consider *decision-time* attacks on machine learning. As we mentioned, prototypical examples of this problem are adversarial evasion of spam, phishing, and malware detectors trained to distinguish between benign and malicious instances, with adversaries manipulating the nature of the objects, such as introducing clever word misspellings or substitutions of code regions, in order to be misclassified as benign.

The key challenge in analyzing robustness of learning approaches to decision-time attacks is in *modeling* these attacks. A good model must account for the central tradeoff faced by an attacker: introducing sufficient manipulation into the object (e.g., malware) to be misclassified as benign as possible (we make this more precise below), while limiting changes to maintain malicious functionality and minimize effort. In this chapter we introduce common mathematical models of decision-time attacks, and associated algorithmic approaches for solving the nontrivial problem of computing an optimal (or near-optimal) attack. In particular, we break the modeling discussion into two subsections based on the *information* dimension in our attack taxonomy: we start with white-box attacks, and discuss black-box attacks thereafter. Moreover, we describe common models of white-box attacks for many of the major classes of learning algorithms: starting with the decision-time attack on binary classifiers (commonly known as the *evasion attack*), we generalize the model to multiclass classifiers, then proceed to describe attacks on anomaly detection, clustering, regression, and, finally, reinforcement learning.

We begin by describing several examples of evasion attacks from the cybersecurity literature to illustrate the challenges involved in devising and executing such attacks in practice, challenges which are typically abstracted away in mathematical models of decision-time attacks that we turn to thereafter.

4.1 EXAMPLES OF EVASION ATTACKS ON MACHINE LEARNING MODELS

We'll begin with several examples of evasion attacks—an important subclass of decision-time attacks—which come from the cybersecurity literature. In an evasion attack, the learned model is used to detect malicious behavior, such as an intrusion or a malicious executable, and the attacker aims to change the characteristics of the attack to remain undetected.

The first example is the *polymorphic blending attack*, the target of which is an intrusion detection system based on statistical anomaly detection. The second set of examples concerns evasion attacks on Portable Document Format (PDF) malware classifiers. We describe several instances of such attacks: one which can be viewed as a *mimircy* attack, as it simply attempts to introduce benign characteristics into a malicious file (rather than removing malicious-looking aspects), and another which automates the attack and can both add and remove PDF objects.

4.1.1   ATTACKS ON ANOMALY DETECTION: POLYMORPHIC BLENDING

Malware polymorphism is a long-standing issue in signature-based malware and intrusion detection systems (IDSs). Since malware signatures tend to be somewhat rigid (looking for an exact, or close match), attacks routinely look to make small modifications to malware code or packaging (such as packing and obfuscation) to significantly modify a hash-based signature. While such attack principles are clearly examples of evasion, they are not pertinent to machine learning. However, anomaly detection has been proposed as a way to resolve the issue of polymorphic malware, since statistical properties of such instances tend to remain quite unlike typical network traffic. Specifically, an anomaly detector in the IDS use case would flag malware as it is being trasmitted over the network.

A generic way to think of anomaly detection systems is to translate entities being modeled (such as network traffic in IDS) into a numeric feature vector, say, x. For example, a common approach is to use n-grams, or features which correspond to sequences of n consecutive bytes, as described in Section 2.2.4. The feature vector corresponding to a particular packet could then be a series of frequencies of each possible n-gram appearing in the packet, or a binary vector indicating for each n-gram whether it appears in the packet. In any case, we can then obtain a data set of "normal" traffic, and model the distribution of the associated feature vectors. If we add a likelihood threshold, we can flag any packets with likelihood (given the distribution of normal data) below the threshold.

A simple (and rather common) way to instantiate such a scheme is to take the mean of the normal traffic feature vector, μ, and impose a threshold on the mean so that any $x : \|x - \mu\| > r$ is flagged as abnormal (with r chosen to achieve a target low false-positive rate). This then becomes an instance of centroid-based anomaly detection we discussed earlier in Section 2.2.4.

Now we can describe (briefly) polymorphic blending attacks. The goal of these is to create polymorphic instances of malware (viewed as a sequence of packets), *but also to achieve the feature representation of the malware which is as close to normal as possible.* Moreover, this needs to be done without risking any deleterious effects on malicious functionality. One way to accomplish this is by a combination of encryption, which substitutes characters common in normal packets for those which are not, decryption, which can ensure that only normal n-grams are used by storing a reverse mapping array with the ith entry having the normal character corresponding to the ith attack character, and padding, which adds more normal bytes to the packet to make it appear even more similar to a normal profile. At execution time, the decryptor then removes the padding, and decrypts the attack packet. Such an attack on anomaly-based IDS was described and evaluated in detail by Fogla et al. [2006].

4.1.2   ATTACKS ON PDF MALWARE CLASSIFIERS

In order to explain PDF malware classification and associated attacks, we first take a brief detour into PDF document structure.

PDF Structure   The PDF is an open standard format used to present content and layout on different platforms. A PDF file structure consists of four parts: *header*, *body*, *cross-reference table* (CRT), and *trailer*. The header contains information such as the format version. The body is the most important element of a PDF file, which comprises multiple objects that constitute the content of the file. Each object in a PDF can be one of eight basic types: Boolean, Numeric, String, Null, Name, Array, Dictionary, and Stream. Moreover, objects can be referenced from other objects via indirect references. There are also other types of objects, such as *JavaScript* which contains executable JavaScript code. The CRT indexes objects in the body, while the trailer points to the CRT. The PDF file is parsed starting with the trailer, which contains the location of the CRT, and then jumps directly to it, proceeding to parse the body of the PDF file using object location information in the CRT.

PDFRate PDF Malware Detector   The PDFRate classifier, developed by Smutz and Stavrou [2012], uses a random forest algorithm, and employs PDF *metadata* and *content* features to categorize benign and malicious PDF files. The metadata features include the size of a file, author name, and creation date, while content-based features include position and counts of specific keywords. Content-based features are extracted by using regular expressions. The features of PDFRate are detailed by Smutz and Stavrou [2012].

Example Attacks on PDFRate Malware Detector   We now briefly describe two attacks on PDFRate. The regular expressions used by PDFRate to generate its features parse the PDF file linearly from beginning to end to extract each feature. In some cases, when features are based on values of a particular PDF object, such as "Author," repeated entries are handled by ignoring all but the last value appearing in the file.

One attack on PDFRate, described by Šrndic and Laskov [2014], involves adding content to a malicious PDF file to make it appear benign to the classifier. This attack leverages the *semantic gap* between PDF readers and the linear file processing performed by the classifier. Specifically, PDF readers following specification begin by reading the trailer, and then jump directly to the CRT. Consequently, any content added *between* the CRT and the trailer would be ignored by a PDF reader, but would still be used to construct features using the PDFRate feature extraction mechanism. While this attack cannot modify all features used by PDFRate, it can change a large number of them. In order to determine which content to add, this attack suggests two approaches: a *mimicry* attack and an attack based on gradient descent in feature space. In the mimicry attack, features of a malicious PDF that can be modified through the attack are transformed to mimic a target benign PDF file in feature space, and then content is added to the malicious PDF to transform one feature at a time to the target value in feature space (or as close to it as feasible). The gradient descent attack optimizes a weighted sum of classification score for a differentiable proxy classifier (e.g., a support vector machine) and an estimated density of benign files. The content is then added to the PDF to match the resulting "optimal" feature vector as close as possible.

Another attack on PDFRate, *EvadeML*, described by Xu et al. [2016], uses genetic programming to directly modify objects in a malicious PDF. EvadeML starts with a malicious PDF which is correctly classified as malicious and aims to produce evasive variants which have the same malicious behavior but are classified as benign. It assumes that the adversary has no internal information about the target classifier, including features, training data, or the classification algorithm. Rather, the adversary has black-box access to the target classifier, and it can repeatedly submit PDF files to get corresponding classification scores. Based on the scores, the adversary can adapt its strategy to craft evasive variants.

EvadeML employs genetic programming (GP) to search the space of possible PDF instances to find ones that evade the classifier while maintaining malicious features. First, an initial population is produced by randomly manipulating a malicious seed. As the seed contains multiple PDF objects, each object is set to be a target and mutated with exogenously specified probability. The mutation is either a deletion, an insertion, or a swap operation. A deletion operation deletes a target object from the seed malicious PDF file. An insertion operation inserts an object from external benign PDF files (also provided exogenously) after the target object. EvadeML uses three most benignly scoring PDF files for this purpose. A swap operation replaces the entry of the target object with that of another object in the external benign PDFs.

After the population is initialized, each variant is assessed by the Cuckoo sandbox [Guarnieri et al., 2012] and the target classifier to evaluate its fitness. The sandbox is used to determine if a variant preserves malicious behavior. It opens and reads the variant PDF in a virtual machine and detects malicious behaviors such as API or network anomalies, by detecting malware signatures. The target classifier (PDFRate, for example) provides a classification score for each variant. If the score is above a threshold, then the variant is classified as malicious. Otherwise, it is classified as a benign PDF. If a variant is classified as benign but displays malicious behavior, or if GP reaches the maximum number of generations, then GP terminates with the variant achieving the best fitness score and the corresponding mutation trace is stored in a pool for future population initialization. Otherwise, a subset of the population is selected for the next generation based on their fitness evaluation. Afterward, the variants selected are randomly manipulated to generate the next generation of the population.

The attacks we described have been demonstrated to be remarkably successful. For example, EvadeML was reported to have 100% evasion success rate [Xu et al., 2016].

4.2 MODELING DECISION-TIME ATTACKS

In order to understand decision-time attacks fundamentally and to allow us to reason about these attacks in general, a number of attempts have been made to model them. Before getting into the mathematical details of several natural models of these attacks, we first describe these conceptually, clarifying some terminology in the process.

An important aspect of a decision-time attack on machine learning, such as the adversarial evasion attack discussed below, is that it is an attack on the machine learning *model*, and **not** on the algorithm. For example, both a linear support vector machine and the perceptron algorithms yield a linear classifier, $f(x) = \text{sgn}(w^T x)$, with feature weights w. From the perspective of a decision-time attack, we only care about the end result, $f(x)$, and not which algorithm produced it. This is not to say that the learning algorithm is irrelevant to robustness of learning to such attacks; rather, one can claim that a particular algorithm *tends to generate* more robust *models* than another algorithm. However, for the purposes of discussing attacks, only the structure of the *model* is relevant.

In a prototypical decision-time attack, an adversary is associated with a particular behavior (e.g., a sequence of commands) or object (e.g., malware) which is being labeled by the learned model as malicious and is thereby prevented from achieving its goal. In response, the adversary makes modifications to said behavior or object aiming to accomplish two objectives: (a) fulfill a malicious objective, such as compromising a host, and (b) significantly reduce the likelihood of being flagged as malicious by the learned model. A related secondary objective is for the attacker to minimize the amount of effort spent devising a successful attack.

To get some intuition about decision-time attacks, consider the following simple example.

Example 4.1   Consider the following example of *adversarial evasion*, a decision-time attack on a binary classifier which flags an instance (say, spam) as malicious or benign. In our example, there is a single feature, which we simply call x. We use a score-based classifier (see Section 2.1.2), $f(x) = \text{sgn}\{g(x)\}$, where $g(x) = 2x - 1$. In other words, an instance x is classified as spam if $g(x) \geq 0$, or $x \geq 0.5$, and non-spam otherwise. We visualize this in Figure 4.1, where the dashed horizontal line represents the $g(x) = 0$ threshold for classifying an instance as malicious vs. benign.

Now, suppose that the spammer created a spam email which is represented by a feature $x_{spam} = 0.7$. The associated $g(x_{spam}) = 0.4 > 0$, as indicated by the heavy red lines in Figure 4.1. In the evasion attack, the spammer would change the spam email so that its corresponding numerical feature x' drops below 0.5, which will ensure that the resulting $g(x') < 0$ (light red lines in Figure 4.1). In other words, the spam email with feature x' is now classified as non-spam.

In the remainder of this chapter we describe how decision-time attacks are commonly modeled and analyzed mathematically.

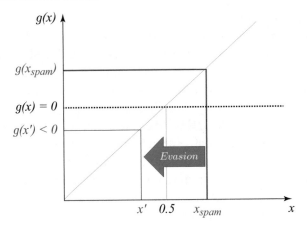

Figure 4.1: Illustration of an evasion attack in Example 4.1.

4.3 WHITE-BOX DECISION-TIME ATTACKS

A major challenge in modeling attacks is the question of what information about the learned model the adversary possesses. We defer discussion of this for now and assume that the adversary knows the model being attacked; that is, we start by considering *white-box* attacks.

4.3.1 ATTACKS ON BINARY CLASSIFIERS: ADVERSARIAL CLASSIFIER EVASION

A common abstraction of white-box evasion attacks on binary classifiers begins with three constructs. The first is the classifier, $f(x) = \text{sgn}(g(x))$ for some scoring function $g(x)$. The second construct is an adversarial feature vector x^A corresponding to the feature characteristics of the behavior, or object, that the adversary wishes to use. Henceforth, we call x^A the *ideal* instance, in the sense that this is the attack vector which would be used by the attacker if it were not flagged as malicious by the classifier $f(x)$. How do we know what x^A is? In practice, we take these to be examples of previously observed attacks, and would consider how each of these would evade the classifier. The third construct is the *cost function*, $c(x, x^A)$ which assigns a cost to an attack characterized by a feature vector x. This cost is meant to capture the difficulty of modifying the ideal instance x^A into x, which may stem from any source, including, crucially, any degradation of malicious functionality. Consequently, it is natural that $c(x^A, x^A) = 0$ (the adversary incurs no cost for leaving the point unperturbed), and that cost of x would increase with distance from x^A in feature space. Evasion attacks will, as we will see presently, aim to balance two considerations: appearing benign to the classifier, either captured by $g(x) \leq 0$ or $f(x) = -1$, and minimizing cost $c(x, x^A)$. Notice that adversarial evasion is fundamentally a *targeted* attack in our terminology, since the adversary wishes to have specific malicious instances classified as benign.

Attack Models   The most common way to model evasion costs is by using l_p distance,

$$c(x, x^A) = \|x - x^A\|_p, \tag{4.1}$$

or its weighted generalization

$$c(x, x^A) = \sum_j \alpha_j |x_j - x_j^A|, \tag{4.2}$$

where the weight α_j aims to capture the difficulty of changing a feature j. Most commonly, l_0, l_1, l_2, or l_∞ norms are used (with $p = 0, 1, 2$, and ∞, respectively). We call all such variations of cost functions *distance-based*, as they are based on a measure of distance (l_p norm) in feature space between a modified and original feature vector. An interesting variation on the distance-based cost function is a *separable* cost function:

$$c(x, x^A) = \max\{0, c_1(x) - c_2(x^A)\}, \tag{4.3}$$

which assumes, roughly, that the cost incurred for generating an evasive instance x is independent of the target instance x^A (modulo the constraint the the final cost is non-negative).

A limitation of distance-based cost functions is that they fail to capture an important feature of real attacks: substitutability or equivalence among attack features. Take spam detection as an example. One common way to construct features of spam email text is by using a bag-of-words representation, where each feature corresponds to the incidence of a particular word. In the simplest case, a binary feature representation would simply have a 1 whenever the corresponding word occurs in the email, and 0 otherwise. An attacker may substitute one word for another, say, using a synonym, without significantly changing the semantics of the message. It is reasonable that such *feature cross-substitution* attacks incur a zero cost. To model this, suppose each feature j has an equivalence class F_j of other features which can be "freely" used as substitutes for j. The cost function can then be represented as

$$c(x, x^A) = \sum_j \min_{k \in F_j | x_j^A \oplus x_k = 1} \alpha_j |x_k - x_j^A|, \tag{4.4}$$

where $\oplus$ is the exclusive-or, so that $x_j^A \oplus x_k = 1$ ensures that we only substitute between different features rather than simply add features.

Whatever cost function one uses, the next question is how to represent the tradeoff faced by the attacker between appearing benign to the classifier, and minimizing evasion cost. Perhaps the most intuitive way to represent this is through the following optimization problem:

$$\min_x [\min\{g(x), 0\} + \lambda c(x, x^A)], \tag{4.5}$$

where λ is a parameter trading off the relative importance of appearing more benign and incurring evasion costs. Notice that the $\min\{g(x), 0\}$ implies that the attacker obtains zero utility if they are classified as malicious, but benefits from looking more benign (i.e., having a smaller $g(x)$). This reflects typical evasion attacks in the security literature, where the attacks explicitly aspire to appear as benign to the classifier as possible. However, this term also makes the optimization non-convex even if the feature space and $g(x)$ are convex. A natural convex relaxation is the following alternative objective for the attacker:

$$\min_x [g(x) + \lambda c(x, x^A)]. \tag{4.6}$$

Another modeling approach is to assume that the attacker cares solely about appearing benign, without any concern about degree of benignness. This can be captured in a few essentially equivalent models. One is the following optimization problem:

$$\min_x \quad c(x, x^A) \tag{4.7a}$$
$$\text{s.t.:} \quad f(x) = -1. \tag{4.7b}$$

While this model is intuitive, it has the property that the attacker can *always* succeed as long as there exists some feature vector x which is classified by $f(x)$ as benign. Consequently, this model is most useful in the analysis of classifier vulnerability in a modified form where we also impose a cost budget constraint, C. More precisely, suppose that x^* solves Problem 4.7. The attacker's decision rule for choosing an attack feature vector x_{new} is then

$$x_{new} = \begin{cases} x^* & \text{if} \quad c(x^*, x^A) \le C \\ x^A & \text{o.w.} \end{cases} \tag{4.8}$$

This is essentially the model used by Dalvi et al. [2004], who term the associated problem *minimum cost camouflage*.

An alternative way to look at this is to have the attacker solve the following optimization problem:

$$\min_x [f(x) + \lambda c(x, x^A)] \tag{4.9}$$

(note the replacement of $g(x)$, which is real-valued, with $f(x)$, which is binary). This problem yields an equivalent decision rule for the attacker as the one we had just described, with the budget constraint $C = 2/\lambda$.

Yet another variation on this theme is the following optimization problem for the attacker:

$$\min_x \quad g(x) \tag{4.10a}$$
$$\text{s.t.:} \quad c(x, x^A) \le C. \tag{4.10b}$$

This also imposes a cost budget constraint on the attacker, but rather than focusing on minimizing the evasion cost, attempts to make an instance look as benign as possible.

Several interesting variations on this last optimization framework consider a somewhat more involved set of constraints on feasible modifications that the adversary can make, and replace the objective with the defender's loss, $l(g(x))$. The first example, *free-range attack*, assumes that the adversary has the freedom to move data anywhere in the feature space. The only knowledge the adversary needs is the valid range of each feature. Let x_j^{max} and x_j^{min} be the largest and the smallest values that the jth feature can take. An attack instance x is then bounded as follows:

$$C_f x_j^{min} \leq x_{ij} \leq C_f x_j^{max}, \forall j, \tag{4.11}$$

where $C_f \in [0, 1]$ controls the aggressiveness of attacks: $C_f = 0$ implies that no attack is possible, while $C_f = 1$ corresponds to the most aggressive attacks involving the widest range of permitted data movement.

The second example, *restrained attack*, attempts to move the initial malicious feature vector x^A toward a particular target x^t. The adversary can set x^t as a prototypical benign feature vector, such as the estimated centroid of innocuous data, a data point sampled from the observed innocuous data, or an artificial data point generated from the estimated innocuous data distribution.

In most cases, the adversary cannot change x^A to x^t as desired since these would compromise malicious value of the attack. To capture this intuition, the restrained attack imposes several constraints that the new evasive instances x must satisfy. First,

$$(x - x^A) \circ (x^t - x^A) \geq 0.$$

This ensures that modifications are in the same direction from x^A as the target. Furthermore, this attack places an upper bound on the amount of displacement for attribute j as follows:

$$|x_j - x_j^A| \leq C_\xi \left(1 - C_\delta \frac{|x_j^t - x_j^A|}{|x_j^A| + |x_j^t|} \right) |x_{ij}^t - x_{ij}|, \tag{4.12}$$

where $C_\delta, C_\xi \in [0, 1]$ model the relative loss of malicious utility in proportion to displacement of original feature vector toward the target. Jointly, these parameters govern how aggressive the attack can be. The term $1 - C_\delta \frac{|x_{ij}^t - x_{ij}|}{|x_{ij}| + |x_{ij}^t|}$ bounds the magnitude of the evasion attack relative to x^A in terms of the original distance between the target and ideal instances: the farther apart these are, the smaller proportion of $|x_j^t - x_j^A|$ can be affected by the attacker.

Computing Optimal Attacks   Now that we have defined a number of stylized models of adversary's objectives in evading a classifier, the next question is: how can we actually solve these optimization problems? We now address this question.

First, observe that if $g(x)$ and $c(x, x^A)$ are convex in x, and $x \in \mathbb{R}^m$ are continuous, almost all of the above formulations of the attacker's decision problem are convex, and can therefore be solved using standard convex optimization techniques. As a simple example, suppose that $g(x) = w^T x$ and $c(x, x^A) = \|x - x^A\|_2^2$ (the squared l_2 norm). Optimizing Problem (4.6) would then yield a closed-form solution

$$x^* = x^A - \frac{2}{\lambda} w.$$

More generally, however, the optimization problems may be non-convex. If we assume that the feature space is real-valued and the objective is sufficiently smooth (e.g., has a differentiable cost function and $g(x)$), one of the most basic techniques for solving these to obtain a locally optimal solution is Gradient descent. To illustrate, suppose the attacker aims to optimize Problem (4.6). If the gradient of the objective is

$$G(x) = \nabla_x g(x) + \nabla_x c(x, x^A),$$

the gradient descent procedure would iteratively apply the following update steps:

$$x_{t+1} = x_t - \beta_t G(x), \tag{4.13}$$

where β_t is the update step. If $g(x)$ and $c(x, x^A)$ are sufficiently smooth, second-order methods, such as Newton-Rhaphson, would be effective as well [Nocedal and Wright, 2006]. Of course, both of these approaches would yield an optimal solution if the attacker's problem is convex.

However, commonly the feature space is discrete, or even binary. A simple general-purpose approach for tackling such problems is a set of methods we collectively call *coordinate greedy (CG)*. In CG, one first chooses a random order over the features, and then iteratively attempts to change one feature at a time from the set of possibilities (we assume that this set is finite, as is typically the case when features are discrete), choosing the best value for this feature while keeping all others fixed. The process stops either after a fixed number of iterations, or when it converges to a locally optimal solution.

Often, even a problem with discrete features can have enough special structure to admit effective global optimization approaches. An example is the approach by Dalvi et al. [2004], who describe a minimum cost camouflage attack, which can be optimally computed using an integer linear program. To simplify discussion, we assume in this case that the feature space is binary (the approach is more general in the original paper). Their attack is specific to a Naive Bayes (NB) classifier (actually, its cost-sensitive generalization). The NB classifier computes the probabilities $p_+(x) = \Pr\{y = +1|x\}$ and $p_-(x) = 1 - p_+(x)$, and predicts $+1$ iff

$$\log(p_+(x)) - \log(p_-(x)) > r$$

for some threshold r (which can be obtained, for example, by considering relative importance of false positives and false negatives). Since for the NB classifier $\Pr\{y = +1|x\} = \Pr\{y = +1\} \prod_j \Pr\{x_j|y = +1\}$, and similarly $\Pr\{y = -1|x\} = \Pr\{y = -1\} \prod_j \Pr\{x_j|y = -1\}$, we can rewrite the decision equivalently as

$$\log \Pr\{y = +1\} + \sum_j \log \Pr\{x_j|y = +1\} - \log \Pr\{y = -1\} - \sum_j \log \Pr\{x_j|y = -1\} > r$$

or

$$\sum_j \left[\log \Pr\{x_j | y = +1\} - \log \Pr\{x_j | y = -1\}\right] > r',$$ (4.14)

where

$$r' = r - (\log \Pr\{y = +1\} - \log \Pr\{y = -1\}).$$ (4.15)

Let us define $L_j(x_j) = \log \Pr\{x_j | y = +1\} - \log \Pr\{x_j | y = -1\}$, $L(x) = \sum_j L_j(x_j)$, and $\text{gap}(x) = L(x) - r'$. $\text{gap}(x)$ is significant as it captures the minimal transformation of the log odds to yield a negative classification of the feature vector, which is to say, to get an instance to be classified as benign rather than malicious. A crucial observation is that modifications to the features impact classification decision independently through $L_j(x_j)$. Moreover, we can define the net impact of flipping the feature x_j (changing it from 1 to 0, or from 0 to 1) as

$$\Delta_j(x_j) = L_j(1 - x_j) - L_j(x_j).$$

The attacker's goal is, thus, to induce a total change to an original instance x classified as malicious that exceeds $\text{gap}(x)$.

Now we can formulate the attacker's optimization problem as the following integer linear program, in which z_j are binary decision variables determining whether a feature j is modified:

$$\min_z \sum_j z_j$$ (4.16a)

$$\text{s.t.} : \quad \sum_j \Delta_j(x_j^A) z_j \geq \text{gap}(x); \quad z_j \in \{0, 1\},$$ (4.16b)

where x_j^A is the value of feature j in the original "ideal" adversarial instance x^A. This is just a variation of the Knapsack problem [Martello and Toth, 1990], and can be solved extremely fast in practice, even though it is NP-Hard in theory Kellerer et al. [2004]. Finally, once the minimum cost camouflage is computed, the result is compared to the adversary's cost budget. The adversary only implements the associated manipulation x' if the modification from the original ideal instance is below the cost budget.

Given the difficulty of achieving an optimal solution for the general adversarial optimization problems, an alternative approach is to aim for algorithms achieving good worst-case approximation guarantees. There have been relatively few examples of this, but a noteworthy case is an algorithm by Lowd and Meek [2005a] for solving Problem (4.7) with a uniform l_1 cost function (i.e., $\alpha_i = 1$ for all i in cost function (4.2)). The algorithm starts with an arbitrary benign instance, $x = x^-$. It then repeats two loops until no further changes are possible: the first loop attempts to flip each feature in x (the current feature vector) which is different from the ideal instance x^A; the second loop attempts to replace a pair of features which are different from x^A with some other feature which is currently identical in x and x^A (but would thereby become

---

Algorithm 4.1 Lowd and Meek Approximation Algorithm for Evading Linear Classifiers

---

$x \leftarrow x^-$
repeat
  $x_{last} \leftarrow x$
  for each feature $j \in C_x$ **do**
    flip x_j
    if $f(x) = +1$ **then**
      flip x_j
    end if
  end for
  for each triple of features $j, k \in C_x, l \notin C_x$ **do**
    flip x_j, x_k, x_l
    if $f(x) = +1$ **then**
      flip x_j, x_k, x_l
    end if
  end for
until $x_{last} = x$
return x

---

different). In either loop, a change is implemented iff the new feature vector is still classified as benign. Notice that each such potential change in either loop would reduce the cost by 1. The full algorithm is given in Algorithm 4.1, where C_x is the set of features which are different between a feature vector x and the ideal instance x^A. Lowd and Meek show that this algorithm approximates the optimal solution to a factor of 2.

4.3.2   DECISION-TIME ATTACKS ON MULTICLASS CLASSIFIERS

Having introduced the basic concepts of decision-time attacks in the context of binary classifiers, we now procede to generalize these to attacks on multiclass classifiers.

Starting with targeted attacks, suppose that the attacker aims to transform the ideal instance x^A so that it is labeled as a target class t. A natural, and very general, model of such attacks is the following optimization problem:

$$\min_x c(x, x^A) \quad \text{s.t.:} \quad f(x) = t, \tag{4.17}$$

where we can additionally impose a cost budget constraint as we had done above. If we wish instead to consider a reliability attack, we can replace the constraint in model (4.17) with $f(x) \neq y$, where y is the correct label.

Typically, however, we have more structure on the multiclass classifier, as it can be commonly represented as

$$f(x) = \arg\max_y g_y(x) \tag{4.18}$$

for some score function $g_y(x)$ (observe that this is a direct generalization of score-based binary classification, where $f(x) = \text{sgn}\{g(x)\}$). In that case, we can transform formulation (4.6) into

$$\max_x [g_t(x) - \lambda c(x, x^A)] \tag{4.19}$$

for a targeted attack, or

$$\min_x [g_y(x) + \lambda c(x, x^A)] \tag{4.20}$$

for a reliability attack (where y is the correct label for x).

Unfortunately, it turns out that this generalization of adversarial evasion to multiclass classifiers is problematic. To illustrate, consider a targeted attack in which the attacker's goal is to ensure that $f(x) = t$ for some target class t. When we solely maximize $g_t(x)$, however, we may also inadvertently maximize $g_y(x)$ for some other class $y \neq t$, with the end result that $g_t(x^*) < g_y(x^*)$ for the optimal solution x^* of Problem (4.19). In other words, the attacker may fail to achieve its goal of inducing a classification of x^* as a target class t.

To address this issue, observe that to obtain $f(x) = t$ the attacker needs the following condition to hold:

$$g_t(x) \geq g_y(x) \quad \forall\, y \neq t.$$

We can increase robustness of targeted attacks by adding a safety margin of γ into this condition, obtaining

$$g_t(x) - \gamma \geq g_y(x) \quad \forall\, y \neq t.$$

Rewriting this, we have the condition that

$$\max_{y \neq t} g_y(x) - g_t(x) \leq -\gamma. \tag{4.21}$$

Carlini and Wagner [2017] suggest replacing $g_t(x)$ in the objective of Problem (4.19) with a function $h(x; t)$ which has the property that $h(x; t) \leq -\gamma$ iff Condition (4.21) is satisfied (that is, iff $f(x) = t$, with the added margin γ). One of these which performed particularly well in their experiments is

$$h(x; t) = \max\{-\gamma, \max_{y \neq t} g_y(x) - g_t(x)\}. \tag{4.22}$$

We consequently rewrite the optimization problem for the targeted attack as

$$\min_x [h(x; t) + \lambda c(x, x^A)]. \tag{4.23}$$

An analogous transformation can be devised for reliability attacks. While the resulting optimization problems are typically non-convex, standard methods, such as gradient descent or local search, can be applied to solve them [Hoos and Stützle, 2004, Nocedal and Wright, 2006].

Attacks on multiclass classifiers have become particularly important with the study of robustness of deep learning algorithms. We devote Chapter 8 entirely to this topic.

4.3.3   DECISION-TIME ATTACKS ON ANOMALY DETECTORS

Some of the well-known evasion attacks have been deployed not against classifiers, but against anomaly detection systems. While there are typically major differences in the specific approaches for anomaly detection and classification (the former being unsupervised, while the latter supervised), it turns out that decision-time attacks on anomaly detectors are essentially identical to evasion attacks on binary classifiers.

To appreciate this, consider the centroid-based anomaly detector with a given μ. As discussed in Section 2.2.4, a feature vector x is viewed as anomalous if $\|x - \mu\| \geq r$. Now, define $g(x) = \|x - \mu\| - r$. We can see that an instance x is classified as anomalous when $f(x) = \text{sgn}\{g(x)\} = +1$ or, equivalently, when $g(x) \geq 0$. Consequently, all the attacks we discussed for binary classifiers are directly applicable here. Similarly, for PCA-based anomaly detectors, we can define

$$g(x) = \|x_e\| - r = \|(\mathbb{I} - \mathbf{V}\mathbf{V}^T)x\| - r, \tag{4.24}$$

where $\mathbf{V}$ (the matrix of eigenvectors produced by PCA) is now given, and again apply standard techniques for binary classifier evasion.

4.3.4   DECISION-TIME ATTACKS ON CLUSTERING MODELS

Just as decision-time attacks on anomaly detectors are conceptually equivalent to evasion attacks on binary classifiers, a natural class of decision-time attacks on clustering is equivalent to attacks on multiclass classifiers.

Let us generically designate a clustering model by partitioning the entire feature space $\mathcal{X}$ into K subsets $\{S_1, \ldots, S_K\}$ corresponding to K clusters. An arbitrary feature vector x then belongs to a cluster k iff $x \in S_k$. Often, such a cluster assignment can be represented as $k \in \arg\max_y g_y(x)$ for some $g_y(x)$. For example, suppose that the assignment is based on l_p distance from a cluster mean, with $\{\mu_1, \ldots, \mu_K\}$ being a collection of cluster means; the common k-means clustering approach is a special case with $p = 2$. Then $k \in \arg\min_y \|x - \mu_y\|_p = \arg\max_y \|x - \mu_y\|_p^{-1}$. In other words, $g_y(x) = \|x - \mu_y\|_p^{-1}$.

With this in mind, we can now define targeted and reliability attacks on clustering as follows. In a targeted attack, the adversary aims to ensure that the instance x^A is miscategorized to belong to a target cluster t with an associated cluster mean μ_t.[1] Similarly, we can define

---

[1] There may be a question here that in clustering the specific cluster identity is not meaningful. However, note that once a clustering model has been produced, it induces meaningful cluster identities, for example, as characterized by cluster means μ_t.

a reliability attack as ensuring that the ideal instance no longer falls into its original (correct) cluster y. Given the definition of $g_y(x)$ above, we can model such attacks on clustering in the same way as we had modeled decision-time attacks on multiclass classifiers.

4.3.5 DECISION-TIME ATTACKS ON REGRESSION MODELS

In decision-time attacks on regression models the attacker, as before, will start with an ideal feature vector x^A, which it aims to transform into another, x', to accomplish either a targeted or reliability attack. In the targeted attack, the attacker has a target regression value t and aims to achieve the predicted value of the regression model, $f(x')$ as close as possible to t. In a reliability attack, on the other hand, the attacker's aim is to cause the prediction $f(x')$ to be far from the correct prediction y.

We illustrate decision-time attacks on regression through the following example.

Example 4.2   Consider a regression function in one variable x shown in dashes in Figure 4.2, with x^A the original ideal attack. Suppose the attacker aims to maximize prediction error after a small modification of x^A. In the figure, this can be accomplished by transforming x^A into x', with $f(x') \gg f(x^A)$ (for example, skewing a stock price predictor to greatly overestimate the price of a stock).

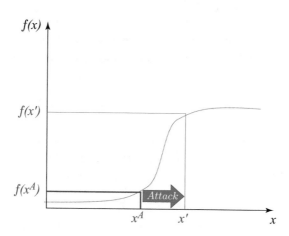

Figure 4.2: Illustration of a decision-time attack on regression in Example 4.2.

We can model targeted attacks by considering an adversarial loss function $l_A(f(x), t)$ which measures, from the adversary's perspective, the error induced by an adversarial instance x with respect to the adversary's prediction target t. This allows us to recapture the canonical tradeoff between achieving adversarial aims and minimizing the amount of manipulation faced by adversaries executing decision-time attacks: the former part of the tradeoff is captured by the loss function $l_A(\cdot)$, while we can capture the latter using the same types of cost functions we discussed above. Thus, the decision-time targeted attack in a regression setting can be modeled using the following optimization problem:

$$\min_{x} l_A(f(x), t) + \lambda c(x, x^A).$$ (4.25)

The analog for a reliability attack is

$$\max_{x} l(f(x), y) - \lambda c(x, x^A),$$ (4.26)

where the attacker now attempts to maximize the learner's loss function $l(f(x), y)$. A common special case for both the loss and the cost function is the squared l_2 norm, i.e., $l_A(f(x), t) = \|f(x) - t\|_2^2$ and $c(x, x^A) = \|x - x^A\|_2^2$.

To offer a concrete illustration, let us consider an attack on a linear regression model [Grosshans et al., 2013]. Suppose that, given a data point (x^A, y) and a linear regression model $f(x) = w^T x$, the attacker aims to bias the prediction toward a target t. If we use (squared) l_2 norm for both the attacker's loss and cost functions, the attacker's optimization problem becomes

$$\min_{x} \quad (w^T x - t)^2 + \lambda \|x - x^A\|_2^2.$$ (4.27)

Now, we can more generically apply this attack to a dataset $\{(x_i^A, y_i)\}$ with an associated vector of corresponding targets t (one for each (x_i^A, y_i)). Let us aggregate all feature vectors into a feature matrix $\mathbf{X}^A$, and all labels into a label vector y. We can write the above optimization problem in matrix form to find the optimal transformation of the feature matrix from $\mathbf{X}^A$ into a new one $\mathbf{X}$:

$$\min_{\mathbf{X}} \quad (\mathbf{X}^A w - t)^T (\mathbf{X}^A w - t) + \lambda \|\mathbf{X} - \mathbf{X}^A\|_F,$$ (4.28)

where $\|\mathbf{X} - \mathbf{X}^A\|_F$ is the Frobenius norm. We can use first-order conditions to characterize the attacker's optimal response to a fixed model parameter vector w; thus, equating the first derivative to zero, we obtain:

$$(\mathbf{X}^* w - t)w^T + \mathbf{X}^* = \mathbf{X}^A,$$ (4.29)

or

$$\mathbf{X}^* = (\lambda I + w w^T)^{-1}(t w^T + \lambda \mathbf{X}^A).$$ (4.30)

Using the Sherman-Morrison formula, we can equivalently write this as

$$\mathbf{X}^* = \mathbf{X}^A - (\lambda + \|w\|_2^2)^{-1}(\mathbf{X}^A w - t)w^T.$$ (4.31)

Each row i in $\mathbf{X}^*$ thus becomes a transformation of an original ideal feature vector x_i^A into an attack x_i^*.

Alfeld et al. [2016] present another interesting variant of attacks on regression models. In particular, they consider attacks on linear autoregressive (AR) models, which are common models used for time series analysis and prediction, for example, in financial markets. Specifically, Alfeld et al. [2016] describe attacks on order-d linear AR models of the form

$$x_j = \sum_{k=1}^{d} w_k x_{j-k}, \qquad (4.32)$$

where $x_j \in \mathbb{R}$ is the scalar observation at time j. In this context, the defender's (learner's) goal is to make predictions at time j for the next h time steps, which can be done by applying Equation (4.32) recursively. The attacker observes, and can modify, the d observations.

In formal notation, suppose that at an arbitrary point in time, the defender's AR(d) model makes use of the previous d observations denoted by $x = (x_{-d}, \ldots, x_{-1})$, where x_{-k} refers to the observation k time steps before the decision time point (i.e., if decision point is time j, then this corresponds to x_{j-k} in Equation (4.32)). The attacker can make modifications to the observed values which inform the AR model, yielding the corrupted observations x'_{-k}. At the same time, modifying a vector of observations incurs a cost, $c(x, x')$.

Just as above, the attacker's goals may be either a targeted attack or an attack on learner reliability. We use the targeted attack as an example, where the attacker has a target vector of predictions $t \in \mathbb{R}^h$ which it wishes to induce. Thus, if x^h are h-step predictions made by the defender, the attacker's corresponding loss is

$$\|x^h - t\|_{\mathbf{W}}^2, \qquad (4.33)$$

where $\|u\|_{\mathbf{W}}^2 = u^T \mathbf{W} u$ is the Mahalanobis norm. Just as in typical models of evasion attacks, it is natural to model cost as a norm, such as squared l_2 distance:

$$c(x, x') = \|x - x'\|_2^2. \qquad (4.34)$$

In this case, we can equivalently represent $x' = x + \delta$ for some attack vector δ, with the cost corresponding to $\|\delta\|_2^2$.

To significantly facilitate analysis of this problem, one can transform it into matrix-vector notation. First, observe that $x_{j+1} = \mathbf{S} x_j$, where $\mathbf{S}$ is the one-step transition matrix:

$$\mathbf{S} = \begin{bmatrix} 0_{h-1} & I_{(h-1)\times(h-1)} \\ 0 & 0_{(h-d-1)\times 1}^T \quad w^T \end{bmatrix}. \qquad (4.35)$$

Consequently, the vector of predictions starting at current time $j = 0$ and through time $j = h-1$ (i.e., over h time steps) is $x_{h-1} = \mathbf{S}^h x_{-1}$ in our notation, where $\mathbf{S}^h$ denotes $\mathbf{S}$ to the power h. Upon adversarial tampering δ, we then obtain the tampered result

$$x'_{h-1} = \mathbf{S}^h (x_{-1} + \mathbf{Z}\delta), \qquad (4.36)$$

where

$$\mathbf{Z} = \begin{bmatrix} 0_{(h-d)\times d} \\ I_{d\times d} \end{bmatrix}. \tag{4.37}$$

Just as discussed above, we can consider a number of attack model variations, with and without constraints. One simple variation which admits a closed form solution is for the attacker to solve the following optimization problem:

$$\min_{\delta} \|\mathbf{S}^h(x_{-1} + \mathbf{Z}\delta) - t\|_W^2 + \lambda\|\delta\|_2^2. \tag{4.38}$$

Observe that this is just the analog of the Problem 4.25 above, with $l_A(f(x), t) = \|\mathbf{S}^h(x_{-1} + \mathbf{Z}\delta) - t\|_W^2$ and $c(x, x^A)$ captured by the squared l_2 norm. The optimal solution is

$$\delta^* = -(\mathbf{Q} + 2\lambda)^{-1}c, \tag{4.39}$$

where

$$\mathbf{Q} = (\mathbf{S}^h\mathbf{Z})^T\mathbf{W}(\mathbf{S}^h\mathbf{Z}) \tag{4.40}$$

and

$$c = \mathbf{Z}^T(\mathbf{S}^h)^T\mathbf{W}^T\mathbf{S}^h x_{-1} - \mathbf{Z}^T(\mathbf{S}^h)^T\mathbf{W}^T t. \tag{4.41}$$

4.3.6   DECISION-TIME ATTACKS ON REINFORCEMENT LEARNING

A decision-time attack in the context of reinforcement learning would fundamentally be an attack on the model learned using RL. Most directly, it is an attack on the *policy*, $\pi(x)$, which maps an arbitrary state x into an action a or, if the policy is randomized, to a probability distribution over actions Δ. The attacker attempts to make modifications to observed state x in a way that leads the defender to make a poor action choice. Suppose that the attacker targets a particular state x which it would transform in an attack to another state x'. Just as in the other attacks, the attacker may have two goals: in a targeted attack, the attacker wants the learned policy π to take a target action $a_t(x)$ (and the target action can be different for different states), while in a reliability attack the attacker aims to cause the learner to take a different action from the one taken by the learned policy $\pi(x)$.

Considering targeted and reliability attacks defined above may appear unnatural for RL: a more natural attack would seem to be to minimize either the direct reward for an action taken in state x, or the expected future stream of rewards. We now show that if the learned model is near-optimal, we can view such attacks equivalently as targeted attacks. Recall that the Q-function $Q(x, a)$ is defined precisely as the expected stream of future rewards in state x if action a is taken, followed subsequently by an optimal policy. If the attacker assumes that, indeed, the defender is playing near-optimally, and if we define $\bar{a} = \arg\min_a Q(x,a)$, then the attack minimizing the expected stream of rewards is equivalent to a targeted attack with the target action $a_t(x) = \bar{a}$. We can deal with the case where the attacker aims to minimize immediate reward similarly.

The next question is: how can we fully model attacks on RL as optimization problems, and how would we solve such problems? We now observe that the targeted and reliability attacks on RL are essentially equivalent to such attacks on multiclass classifiers. To see this, observe that if we assume a greedy policy with respect to a Q-function, then the policy can be defined as

$$\pi(x) = \arg\max_a Q(x,a). \tag{4.42}$$

Now, if we treat a as a predicted class and define $g_a(x) = Q(x,a)$, we can rewrite this as

$$\pi(x) = \arg\max_a g_a(x), \tag{4.43}$$

which is identical to the score-based definition of multiclass classifiers we discussed earlier. Consequently, we can implement targeted and reliability attacks on RL by using identical machinery as for decision-time attacks against multiclass classifiers.

4.4   BLACK-BOX DECISION-TIME ATTACKS

White-box decision-time attacks highlight *potential* vulnerabilities of learning approaches. However, they rely on an assumption that the attacker knows everything there is to know about the system, an assumption which is clearly unrealistic. In order to fully understand vulnerabilities, we now discuss situations where the attacker has only partial information about the learning system, conventionally called *black-box attacks*.

There are two core issues in understanding black-box decision-time attacks: (1) how to categorize partial information the attacker may have about the system and what can be accomplished with this information; and (2) how to model the way that the attacker can obtain information. We address the first question by describing a comprehensive taxonomy of black-box decision-time attacks. We then discuss a natural *query* framework within which an adversary can obtain information about the learning model, focusing our discussion on the adversarial evasion problem.

4.4.1   A TAXONOMY OF BLACK-BOX ATTACKS

As we observed earlier, decision-time attacks are attacks on the learned *model, f*. Consequently, black-box attacks are fundamentally about information that the attacker possesses, or can infer, about the actual model f used by the learner.

In Figure 4.3, we present a visualization of our taxonomy of black-box decision-time attacks. Given that decision-time attacks are attacks on models, the taxonomy is centered around

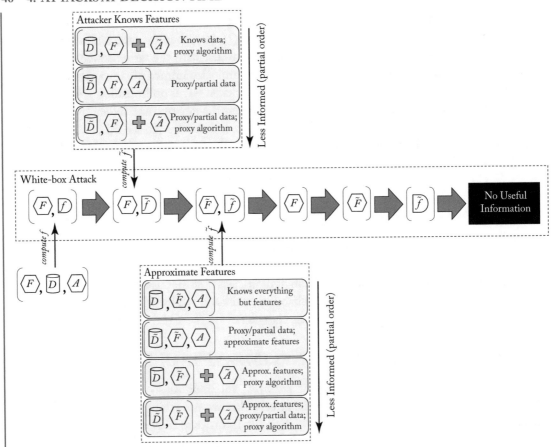

Figure 4.3: A schematic of the hierarchy of black-box decision-time attacks. F denotes the true feature space, f the true model used by the learner, A the actual learning algorithm, and D the dataset on which f was learned by applying A. $\tilde{f}$ is an approximate (proxy) model, $\tilde{F}$ is an approximate feature space, $\tilde{A}$ a proxy algorithm, and $\tilde{D}$ a proxy or partial dataset, with all proxies derived or obtained where corresponding full information is unavailable.

information the attacker may have about the model, including the feature space used. This serves as the organizing principle behind our taxonomy.

In the white-box attack, the attacker knows both the feature space, which we denote by F, and the true model f. Equivalently, it suffices for the attacker to know the dataset D with feature set F on which the learner applied the algorithm A to derive the model f by running A on D (assuming that A is deterministic). On the other hand, if any of F, D, and A are not known exactly, the attacker can only obtain a proxy model $\tilde{f}$ which approximates the true target model f. Additionally, an approximate model $\tilde{f}$ can represent query access to a true model, which we discuss in greater detail below.

Let's start with an attacker who knows the feature set F. If the attacker also knows the dataset, they can use a proxy algorithm to derive a proxy model $\tilde{f}$. Consequently, we arrive at the information state $[F, \tilde{f}]$ (attacker knows features, and has a proxy model). We generally expect this approach to be quite effective in classification learning: as long as both f and $\tilde{f}$ are highly accurate, they are necessarily similar in expectation. Formally, suppose that $h(x)$ is a true function both f and $\tilde{f}(x)$ try to fit using data $\mathcal{D}$, and suppose the error rate of both f and $\tilde{f}$ is bounded by ϵ. Then,

$$
\begin{aligned}
\Pr_x\{f(x) \neq \tilde{f}(x)\} &= \Pr_x\{[f(x) \neq h(x) \wedge \tilde{f}(x) = h(x)] \vee [f(x) = h(x) \wedge \tilde{f}(x) \neq h(x)]\} \\
&\leq \Pr_x\{[f(x) \neq h(x) \wedge \tilde{f}(x) = h(x)]\} + \Pr_x\{[f(x) = h(x) \wedge \tilde{f}(x) \neq h(x)]\} \\
&\leq \Pr_x\{f(x) \neq h(x)\} + \Pr_x\{\tilde{f}(x) \neq h(x)\} \leq 2\epsilon.
\end{aligned}
$$

Now, even if the attacker only has a proxy dataset $\tilde{\mathcal{D}}$, they can still infer a proxy model $\tilde{f}$ by running either the same algorithm as the learner, or a proxy, on $\tilde{\mathcal{D}}$. Thus, we still arrive at the information state $[F, \tilde{f}]$. The success of such an attack now depends largely on how good an approximation $\tilde{\mathcal{D}}$ for the true data $\mathcal{D}$; for example, if they are both large enough, and come from a similar distribution over instances, we expect $\tilde{f}$ to remain a good approximation of model f.[2]

Note that the same ideas can be used even if the attacker does not know the true feature space, but only uses a proxy $\tilde{F}$. Now, whether they know the data or the algorithm, the attacker can still obtain an approximate model $\tilde{f}$, yielding a somewhat lower information state $[\tilde{F}, \tilde{f}]$ than in the scenarios above. However, the difference between knowing F and $\tilde{F}$ is vital: if the proxy feature space is very different from F, the attack may be much less likely to succeed.

Moving further toward the no-information extreme, the attacker may only have knowledge of F, the features. In this case, an attacker may still be able to execute a *mimicry* attack: for example, given a raw malicious instance (such as malware), and a small collection of benign instances, the attacker can attempt to manipulate the malicious instance directly to make its features close to those for benign instances. A mimicry attack can also be executed with a proxy feature set $\tilde{F}$, but here again it is unlikely to succeed unless $\tilde{F}$ is a sufficiently good proxy for F. Finally, the attacker may know nothing about features, but may have query access to the true model—an information state we also denote by $\tilde{f}$. In this case, the attacker may be able to directly manipulate the malicious instance (for example, adding and removing objects from a malicious PDF file), repeatedly querying the model to ascertain whether the modification suffices to bypass the detector. This was the nature of the EvadeML attack [Xu et al., 2016].

---

[2]We note that this formalization only offers intuition about effectiveness of attacks designed against $\tilde{f}$ in succeeding against the true target f. Since the argument is about expectations, it is possible for two functions to be very similar in this sense, but sufficiently different for purposes of attacks.

4.4.2   MODELING ATTACKER INFORMATION ACQUISITION

One of the earliest treatments of black-box attacks suggested a natural query model: the attacker has black-box query access, whereby they can submit feature vectors x as input, and observe the label $f(x)$ assigned by the learner (for example, whether a classifier considers x as malicious). Henceforth, we focus our discussion of query models on the problem of adversarial evasion of binary classifiers.

The most basic algorithmic problem in this query model, proposed by Lowd and Meek [2005a], is to solve (either exactly, or approximately) the optimization problem (4.7) (minimizing weighted l_1 evasion cost, subject to being misclassified as benign) with a polynomial number of queries to the classifier $f(x)$. They term this problem *ACRE Learnability* (ACRE stands for *adversarial classifier reverse engineering*). Lowd and Meek [2005a] show that this is NP-Hard if the feature space is binary even when $f(x)$ is linear. However, Algorithm 4.1 that we described earlier in the context of white-box attacks can actually also be used in this query model, obtaining a 2-approximation. A follow-up study on learnability showed that convex-inducing classifiers are also approximately ACRE learnable, albeit over a continuous feature space [Nelson et al., 2012].

The advantage of the query model is that it does not aim to directly approximate the classifier $f(x)$, but only "asks" a series of specific, but possibly costly questions (hence learnability in terms of the number of queries, which is viewed as the costly operation). An alternative approach one may consider is to first use queries to approximately learn (reverse engineer) $f(x)$, and then solve problem (4.7) (which no longer requires expensive queries). However, even learning linear classifiers is NP-Hard unless the target function is also linear [Hoffgen et al., 1995], so this route appears no less challenging.

Fortunately, from an attacker's perspective the learning problem is very special: in our query model, the labels correspond to actual classification decisions with no noise, and, moreover, a classifier being reverse engineered *has itself been learned*! This property should be sufficient to make reverse engineering easy.

To formalize this intuition, we can appeal to the well-known concept of polynomial learnability. Recall from Chapter 2 that (informally) a hypothesis class $\mathcal{F}$ is learnable if we can compute a nearly-optimal candidate from this class for an arbitrary distribution $\mathcal{P}$ over data. In our context, learning will be performed at two levels: first, by the "defender," who is trying to distinguish between good and bad instances, and second, by an "attacker," who is trying to infer the resulting classifier. We call the *attacker's* learning task the *reverse engineering* problem, with an additional restriction: we require the attacker to be within a small error, γ, with respect to the actual model f used by the defender.

Definition 4.3   We say that a distribution $\mathcal{P}$ over data (x, y) can be efficiently γ-*reverse engineered* using a hypothesis class $\mathcal{F}$ if there is an efficient learning algorithm $L(\cdot)$ for $\mathcal{F}$ with the property that for any $\epsilon, \delta \in (0, 1)$, there exists $m_0(\epsilon, \delta)$, such that for all $m \geq m_0(\epsilon, \delta)$, $\mathrm{Pr}_{z^m \sim \mathcal{P}}\{e(L(z^m)) \leq \gamma + \epsilon\} \geq 1 - \delta$.

As the following result demonstrates, efficient learning directly implies efficient 0-reverse engineering.

Theorem 4.4   *Suppose that $\mathcal{F}$ is polynomially (PAC) learnable, and let $f \in \mathcal{F}$. Then any distribution over (x, y) with $x \sim \mathcal{P}$ for some $\mathcal{P}$ and $y = f(x)$ can be efficiently 0-reverse engineered.*

This result is an immediate consequence of the definition of learnability. Thus, insofar as we view the (empirical) efficiency of the defender's learning algorithm as a practical prerequisite, this result suggests that reverse engineering is easy *in practice*. Moreover, this idea also suggests a general algorithmic approach for reverse engineering classifiers:

1. generate a polynomial number of feature vectors x,

2. query the classifier f for each x generated in step 1; this yields a dataset $\mathcal{D} = \{x_i, y_i\}$, and

3. learn $\tilde{f}$ from $\mathcal{D}$ using the same learning algorithm as applied by the defender.

The key limitation, of course, is that this approach implicitly assumes that the attacker knows the defender's learning algorithm, as well as feature space. We can relax this restriction by using a proxy algorithm $\tilde{A}$ and a proxy feature space $\tilde{F}$ for this procedure, but clearly it may significantly degrade the effectiveness of the approach.

In a variation of this query model the adversary can observe classification scores $g(x)$ for queried instances x rather than only the classification decisions $f(x)$. Since we can translate classification scores into classification directly, all of the results about reverse engineering carry over. In practice, however, the classification scores are typically easier to learn.

4.4.3   ATTACKING USING AN APPROXIMATE MODEL

Even if we are able to obtain a reasonably good estimate of the decision function $f(x)$ or scoring function $g(x)$, it is never *exactly* correct. Let's denote the approximation of the scoring function by $\tilde{g}(x)$. One approach is to simply use either $\tilde{f}(x)$, or the approximate score $\tilde{g}(x)$, in an optimization problem used for white-box attacks. However, an attacker may wish to be somewhat more conservative to ensure that the attack is most likely to succeed. We now describe an approach to explicitly account for such model uncertainty in the context of the special case of adversarial evasion.

In an evasion attack, one way for the attacker to hedge against uncertainty is to incorporate a term which scores similarity of a modified instance x to benign feature vectors. To this end, the following intuition is helpful: we would expect that a successful attack should appear to be reasonably likely to be benign, based on typical benign data. To make this precise, suppose that $p_b(x)$ represents a density function of feature vectors x from the benign class. The attacker's optimization problem can then be modified by including the density term as follows (in the case of Problem (4.6)):

$$\min_x \tilde{g}(x) + \lambda c(x, x^A) - \beta p_b(x).$$ (4.44)

Here, observe that in addition to the conventional tradeoff between attack goal (captured by $\tilde{g}(x)$) and cost (captured by $c(x, x^A)$), we attempt to also maximize the likelihood of x being benign, based on the density of benign data $p_b(x)$.

One may naturally wonder how we can obtain (or approximate) the density function $p_b(x)$. One way, suggested by Biggio et al. [2013], is by applying kernel-based nonparametric density estimation, using a dataset of known benign instances D_-, where

$$p_b(x) = \sum_{i \in D_-} k\left(\frac{x - x_i}{h}\right),$$ (4.45)

where $k(\cdot)$ is the kernel function, and h the kernel smoothness parameter.

4.5   BIBLIOGRAPHICAL NOTES

Important early examples of evasion attacks termed *polymorphic blending attacks* on anomlay detection models are presented by Fogla et al. [2006], and then significantly generalized by Fogla and Lee [2006]. Our discussion of polymorphic blending attacks attacks is based on these efforts. The more recent attack on PDF malware classifiers using genetic programming was described by Xu et al. [2016], which is the source of our discussion on that topic.

The earliest algorithmic treatments of classifier evasion are due to Dalvi et al. [2004] and Lowd and Meek [2005a]. Dalvi et al. [2004] considered both the evasion attack which they termed *minimum cost camouflage*, and the meta-problem of developing a more robust classifier that we tackle in the next chapter. Lowd and Meek [2005a] introduce a number of seminal models and results, including problem (4.7) as a model of the attacker's decision problem, and the ACRE learnability concept. They also propose algorithmic approaches to ACRE learnability for linear classification models, including the two-approximation result for polynomial learnability over a binary feature space. Nelson et al. [2012] extend ACRE learnability results by considering convex-inducing classifiers—that is, classifiers which induce some convex classification region, either for the positive or the negative examples. Our discussion of learning proxy models in a black-box attack setting is closely related to the more recent discussion of black-box attacks and transferability in deep learning [Papernot et al., 2016c].

Barreno et al. [2006] and Nelson et al. [2010] consider models of attacks on machine learning, as well as the issue of "defending" classifiers, or amending standard algorithms to produce classifiers that are more robust to evasion. A number of efforts by Biggio et al. [2013, 2014b] introduce several variations of attack models and associated optimization problems [Zhang et al., 2015], and were the first to develop attacks against nonlinear classification models. The separable cost function was introduced by Hardt et al. [2016]. Li and Vorobeychik [2014] discuss the idea of feature substitution, and introduce a cost function to capture it. They also demonstrate the deleterious impact that feature reduction may have on classifier robustness to evasion attacks, particularly when features can be substituted for one another. Zhou et al. [2012] describe free-range and restrained attack models.

Evasion attacks on multiclass classifiers have been studied largely in the context of deep learning (see, e.g., Carlini and Wagner [2017]), where we borrow some of the modeling ideas). We take these up in greater detail, with more comprehensive bibliographic notes, in Chapter 8. This, incidentally, is where we'll also discuss evasion attacks against reinforcement learning models, which have not seen much work outside of deep reinforcement learning.

There has been relatively little work on evasion attacks against regression models. Our discussion for linear regression follows Grosshans et al. [2013], who derive the attack in the more general case which includes uncertainty about the attacker's cost-sensitive learning model. We omit both the cost-sensitive aspects of the model, and the associated uncertainty, to streamline the discussion. In any case, arguably the more significant aspects that are uncertain to the learner in the context of an evasion attack are (a) how the attacker trades off evasion cost and goals, and (b) what the target is in the targeted attack. To the best of our knowledge, none of these issues have been considered in prior literature. The attack on linear autoregressive models which we describe is due to Alfeld et al. [2016].

Lowd and Meek [2005a], along with foundational modeling ideas, also introduced the first query-based black-box attack on classifiers, and follow-up work on ACRE learnability takes the same modeling approach. Vorobeychik and Li [2014] discuss classifier reverse engineering more generally, and the result about 0-reverse engineering of efficiently learned classifiers is due to them. The definition of learnability in that section is adapted from Anthony and Bartlett [2009].

C H A P T E R   5

Defending Against Decision-Time Attacks

In the previous chapter we discussed a number of classes of decision-time attacks on machine learning models. In this chapter, we take up the natural follow-up question: how do we defend against such attacks? As most of the literature on robust learning in the presence of decision-time attacks is focused on supervised learning, our discussion will be restricted to this setting. Additionally, we deal with an important special case of such attacks in the context of deep learning separately in Chapter 8.

We begin this chapter by providing a general conceptual foundation for the problem of hardening supervised learning against decision-time attacks. The remainder of the chapter is organized as follows.

1. **Optimal Hardening of Binary Classifiers:** this section addresses the problem of making binary classifiers robust to decision-time (mostly, evasion) attacks, and focuses on several important special cases which can be solved optimally.

2. **Approximately Optimal Classifier Hardening:** here, we present methods which are significantly more scalable, and in some cases much more general, but are only approximately optimal.

3. **Decision Randomization:** this part of the chapter describes how to develop an approach which hardens general classes of binary classifiers through principled randomization.

4. **Hardening Linear Regression:** finally, we briefly describe a method for hardening linear regression against decision-time attacks.

5.1 HARDENING SUPERVISED LEARNING AGAINST DECISION-TIME ATTACKS

Let us first recall the learning objective when we are not concerned with adversarial data manipulation. This goal is to learn a function f with the property that

$$\mathbb{E}_{(x,y)\sim\mathcal{P}}[l(f(x), y)] \leq \mathbb{E}_{(x,y)\sim\mathcal{P}}[l(f'(x), y)] \quad \forall f' \in \mathcal{F},$$

where $\mathcal{P}$ is the unknown distribution over data. The term $\mathbb{E}_{(x,y)\sim\mathcal{P}}[l(f(x), y)]$ is commonly termed *expected risk* of a classifier f, and we denote it by $\mathcal{R}(f)$. In traditional learning, both the training data, as well as future data on which we wish to make predictions, is distributed according to $\mathcal{P}$.

In adversarial supervised learning, we assume that there is a particular way the distribution of future data is modified as compared to training data: for instances in an adversarial target set $S \subseteq \mathcal{X} \times \mathcal{Y}$, the adversary modifies the corresponding feature vectors to cause prediction error (either to match another target label, as in a targeted attack, or to maximize error, as in a reliability attack).

Let's represent decision-time attacks abstractly as a function $\mathcal{A}(x; f)$ which maps feature vectors, along with the learned model, to feature vectors, that is, taking original feature vectors that they would ideally wish to use (as represented by the training data or the distribution $\mathcal{P}$ from which it is drawn), and modifying them into new feature vectors aimed at changing the predicted label of the learned function f. The resulting expected *adversarial empirical risk*, $\mathcal{R}_A(f)$ for a given function f is

$$
\mathcal{R}_A(f) = \mathbb{E}_{(x,y)\sim\mathcal{P}}[l(f(\mathcal{A}(x; f)), y)|(x, y) \in S] \Pr_{(x,y)\sim\mathcal{P}}\{(x, y) \in S\}
$$
$$
+ \mathbb{E}_{(x,y)\sim\mathcal{P}}[l(f(x), y)|(x, y) \notin S] \Pr_{(x,y)\sim\mathcal{P}}\{(x, y) \notin S\}. \tag{5.1}
$$

Notice that we split the adversarial risk function into two parts: one corresponding to adversarial instances, which behaves according to the model we encode by the function $\mathcal{A}$, and another for non-adversarial instances which are unchanged. The goal is then to solve the *adversarial empirical risk minimization* problem

$$
\min_{f \in \mathcal{F}} \mathcal{R}_A(f). \tag{5.2}
$$

Of course, just as in traditional learning, Problem (5.2) cannot be solved without knowing the distribution $\mathcal{P}$. Instead, we assume the existence of a training data set $\mathcal{D} = \{x_i, y_i\}_{i=1}^n$, which we use as a proxy. In particular, we define the *adversarial empirical risk* function $\tilde{\mathcal{R}}_A(f)$ as

$$
\tilde{\mathcal{R}}_A(f) = \sum_{i \in \mathcal{D}:(x_i,y_i)\in S} l(f(\mathcal{A}(x_i; f)), y_i) + \sum_{i \in \mathcal{D}:(x_i,y_i)\notin S} l(f(x_i), y_i). \tag{5.3}
$$

Then, we approximate Equation (5.2) with the following optimization problem:

$$
\tilde{\mathcal{R}}_A^* = \min_{f \in \mathcal{F}} \tilde{\mathcal{R}}_A(f) + \rho(f), \tag{5.4}
$$

where $\rho(f)$ is a standard regularization function as in traditional learning, and $\tilde{\mathcal{R}}_A^*$ is the minimum empirical adversarial risk.

A useful way to view the problem of hardening learning against decision-time attacks is as a *Stackelberg game*. In a Stackelberg game, there are two players: a leader and a follower. The leader moves first, making its strategic decision, which is then observed by (or becomes known to) the follower. The follower then proceeds to make its own choice, at which point the game ends and the payoffs are realized to both players. In our setting, the leader is the learner, whose strategic choice is the model f. The follower is the attacker, who observes f, and chooses an optimal decision-time attack, which we represent by the mapping $\mathcal{A}(x_i; f)$ for each instance $(x_i, y_i) \in S$, where $\{x_i\}$ encode the attacker's original "ideal" feature vectors. The *solution* to the Stackelberg game is a *Stackelberg equilibrium*, in which the combination of $\{f, \{\mathcal{A}(x_i; f)\}_{i|(x_i, y_i) \in S}\}$ are such that

1. $\mathcal{A}(x_i; f)$ is a best response to f for each $i|(x_i, y_i) \in S$ (that is, the attacker is optimizing given x_i and f, per one of the attack models in Chapter 4), and

2. f minimizes the adversarial empirical risk.

The approaches we discuss in this chapter are either computing a Stackelberg equilibrium of this game for some pre-defined attack model and learner, or approximating it.

Example 5.1 Consider a simple one-dimensional feature space $x \in [0, 1]$ and a data set of two examples, $\mathcal{D} = \{(0.25, benign), (0.75, malicious)\}$. Our goal is to find a threshold r on x to robustly distinguish benign and malicious examples. In our case, any $x > r$ is considered malicious, and $x \le r$ is benign. Moreover, for any such threshold, the attacker would aim to minimize the change in x necessary to be classified as benign, subject to a constraint that the total change is at most 0.5 (corresponding to the Equation (4.7) with the additional budget constraint $C = 0.5$). The example is visualized in Figure 5.1. Consider first a natural baseline in which the threshold $r = 0.5$ is chosen to be equidistant from the benign and malicious instance (this is an example of *maximum margin* classification which is at the root of support vector machines Bishop [2011]). Since the attacker can easily change the instance to be just below 0.5, this threshold is not evasion-robust (the attacker can succeed in evading the classifier). However, an alternative threshold $r = 0.25$ is robust: if we assume that the attacker breaks ties in the learner's favor, the attacker cannot successfully evade the classifier since they can change the original feature by at most 0.5, and the learner will have perfect accuracy even after an evasion attack.

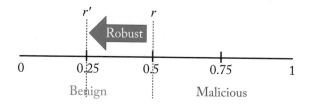

Figure 5.1: Example: evasion-robust binary classification (Example 5.1).

Example 5.1 highlights an interesting issue of tie-breaking. In fact, we use a refinement of the Stackelberg equilibrium concept known as a *Strong Stackelberg equilibrium (SSE)*, where ties among best responses for the attackers are broken in the learner's favor [Tambe, 2011]. The technical reason for this is that an SSE is always guaranteed to exist. A conceptual reason is that it is typically possible for the learner to make small changes in its decision f to make the attacker strictly prefer its equilibrium choice with minimal impact on the defender's loss. For example, if in Example 5.1 the attacker's cost was $C = 0.4$, then the optimal threshold would be $r = 0.35$, and the learner can always make $r = 0.35 - \epsilon$ for an arbitrarily small ϵ and ensure that a successful attack is infeasible without sacrificing accuracy.

An important concern about the Stackelberg game model above is that the adversary may not know the classifier f if it's a black-box attack. However, note that if we are interested in robust learning, it is reasonable to assume that the attack is white-box: first, the learner knows the model f, and second, robustness to a white-box attack implies robustness to a black-box attack. Another caveat is that the actual game is with respect to an unknown distribution of the attackers. In practice, we can use the training dataset as a proxy for this distribution, as is standard in learning.

5.2   OPTIMAL EVASION-ROBUST CLASSIFICATION

We begin discussion of approaches for hardening learning against decision-time attacks by considering optimal evasion-robust binary classification, in the sense of minimizing adversarial empirical risk (AER). In this case it is useful to consider the data set as split into two parts: $\mathcal{D}_+$, which corresponds to instances with $y_i = +1$ (malicious instances), and $\mathcal{D}_-$, corresponding to $y_i = -1$ or $y_i = 0$ (bening instances, depending on how the classes are coded).

5.2.1   OPTIMAL EVASION-ROBUST SPARSE SVM

We start by presenting an optimal evasion-robust binary classification algorithm that can be applied to most of the evasion attack models described in Chapter 4. This algorithm, due to Li and Vorobeychik [2018], is specific to a linear support vector machine binary classifier with l_1 regularization.

Since adversaries correspond to feature vectors x_i which are malicious (and which we interpret as the "ideal" instances x^A of the adversary), we henceforth refer to a given adversarial instance by its index i. We now rewrite the adversarial empirical risk minimization problem in the case of l_1 regularized SVM as a bi-level program in which the learner first chooses the weights w and the attacker modifies malicious instances x_i (i.e., $i \in D_+$) into alternatives, $\tilde{x}_i$, in response:

$$\min_{w} \sum_{i \in \mathcal{D}_-} l_h(-1, w^T x_i) + \sum_{i \in \mathcal{D}_+} l_h(1, w^T \widetilde{x}_i) + \delta||w||_1 \qquad (5.5)$$

$$\text{s.t.:} \quad \forall i \in \mathcal{D}_+,$$

$$z_i = \arg\min_{x|h(x;w)\leq 0} l_A(x, x_i; w)$$

$$\widetilde{x}_i = \begin{cases} z_i & z_i \in \mathcal{C}_i \\ x_i & otherwise, \end{cases}$$

where $l_h(y, w^T x) = \max\{0, 1 - y w^T x\}$ is the hinge loss, $l_A(x,, x_i; w)$ is an adversarial loss function that the attacker wishes to minimize (which may depend on the learning parameters w), subject to constraints $h(x; w) \leq 0$, and $\mathcal{C}_i$ is the set of feasible attacks for an adversarial instance i. An example of the constraints $h(x; w) \leq 0$ is $w^T x \leq 0$, that is, the attacker wishes to ensure that they are classified as benign.

As we can see in the formulation, the decision of the attacker depends on whether or not its budget constraints are satisfied by the optimal adversarial instance (for example, whether it's so far from the original malicious instance that its malicious utility is largely compromised). This is represented by the constraint that $\tilde{x}_i = z_i$ if $z_i \in \mathcal{C}_i$, and otherwise the attacker does not change their original feature vector x_i. A natural example of a budget constraint is $\mathcal{C}_i = \{z | c(z, x_i) \leq C\}$, where C is the attacker's cost budget.

The power of our approach and the formulation (5.5) is that it admits, in principle, *an arbitrary adversarial loss function* $l_A(x, x_i; w)$, and, consequently, an arbitrary cost function. The methods described below will generalize as long as we have an algorithm for optimizing the adversary's loss given a classifier.

In order to solve the optimization problem (5.5) we now describe how to formulate it as a (very large) mathematical program, and then propose several heuristic methods for making it tractable. The first step is to observe that the hinge loss function and $||w||_1$ can both be easily linearized using standard methods. We therefore focus on the more challenging task of expressing the adversarial decision in response to a classification choice w as a collection of linear constraints.

Consider the adversary's core optimization problem which computes

$$z_i = \arg\min_{x|h(x;w)\leq 0} l_A(x, x_i; w) \qquad (5.6)$$

when $z_i \in \mathcal{C}_i$, and with $z_i = x_i$ otherwise.

Now define an auxiliary matrix T in which each column corresponds to a particular attack feature vector x', which we index using variables a; thus, T_{ja} corresponds to the value of feature j in the attack feature vector with index a. Define another auxiliary binary matrix Q where $Q_{ai} = 1$ iff the attack strategy $a \in C$ for the attack instance i.

Next, define a matrix L where L_{ai} is the attacker's loss from the strategy a for an adversarial instance i. Finally, let z_{ai} be a binary variable that selects exactly one feature vector a for the adversarial instance i. First, we must have a constraint that $z_{ai} = 1$ for exactly one strategy a: $\sum_a z_{ai} = 1 \; \forall \; i$. Now, suppose that the strategy a that is selected is the best available option for the attack instance i; it may be below the cost budget, in which case this is the strategy used by the adversary, or above budget, in which case x_i is used. We can calculate the resulting value of $w^T \tilde{x}_i$ inside the loss function corresponding to adversarial instances using

$$w^T \tilde{x}_i = e_i = \sum_a z_{ai} w^T (Q_{ai} T_a + (1 - Q_{ai}) x_i). \tag{5.7}$$

This expression introduces bilinear terms $z_{ai} w^T$, but since z_{ai} are binary, these terms can be linearized using McCormick inequalities [McCormick, 1976].

To ensure that z_{ai} selects the strategy which minimizes the adversary's loss $l_A(\cdot)$ among all feasible options, captured by the matrix L, we can introduce constraints

$$\sum_a z_{ai} L_{ai} \leq L_{a'i} + M(1 - r_{a'}) \quad \forall \; a',$$

where M is a large constant and r_a is an indicator variable which is 1 for an attack strategy a iff $h(T_a; w) \leq 0$ (that is, if feature vector x associated with the attack a, satisfies the constraint $h(x; w) \leq 0$). We can calculate r_a for all a using constraints

$$(1 - 2r_a)h(T_a; w) \leq 0.$$

The resulting full mathematical programming formulation is shown below:

$$\min_{w,z,r} \sum_{i \in D_-} \max\{0, 1 - w^T x_i\} + \sum_{i \in D_+} \max\{0, 1 + e_i\} + \delta \|w\|_1 \tag{5.8a}$$

$$\text{s.t.}: \quad \forall a, i, j : z_{ai}, r_a \in \{0, 1\} \tag{5.8b}$$

$$\sum_a z_i(a) = 1 \tag{5.8c}$$

$$\forall i : e_i = \sum_a m_{ai}(Q_{ai} T_a + (1 - Q_{ai}) x_i) \tag{5.8d}$$

$$\forall a, i, j : -M z_{ai} \leq m_{aij} \leq M z_{ai} \tag{5.8e}$$

$$\forall a, i, j : w_j - M(1 - z_{ai}) \leq m_{aij} \leq w_j + M(1 - z_{ai}) \tag{5.8f}$$

$$\forall a', i : \sum_a z_{ai} L_{ai} \leq L_{a'i} + M(1 - r_{a'}) \tag{5.8g}$$

$$\forall a : (1 - 2r_a)h(T_a; w) \leq 0. \tag{5.8h}$$

Variables m_{ai} allow us to linearize the Constraints (5.7), replacing them with Constraints (5.8d)–(5.8f). Constraint (5.8h) is the only nonlinear constraint remaining, and depends on the specific form of the function $h(T_a; w)$; we deal with it below in two special cases of attack models.

As is, the resulting mathematical program is intractable for two reasons: first, the best response must be computed (using a set of constraints above) for each adversarial instance i, of which there could be many, and second, we need a set of constraints for each feasible attack action (feature vector) x (which we index by a). We can tackle the first problem by clustering the "ideal" attack vectors x_i into a set of clusters and using the mean of each cluster as the ideal attack x^A for the representative attacker. This dramatically reduces the number of adversaries and, therefore, the size of the problem. To tackle the second problem, following Li and Vorobeychik [2018], we can use constraint generation to iteratively add strategies a into the above program by computing optimal, or approximately optimal, attack strategy to add in each iteration.

Algorithm 5.2 AdversarialSparseSVM(X)

$T =$ init() // initial set of attacks
$\mathcal{D}' \leftarrow$ cluster($\mathcal{D}$)
$w_0 \leftarrow$ MILP($\mathcal{D}', T$)
$w \leftarrow w_0$
while T changes **do**
    for $i \in \mathcal{D}'_+$ **do**
        $t =$ computeAttack(x_i, w)
        $T \leftarrow T \cup t$
    end for
    $w \leftarrow$ MILP($\mathcal{D}', T$)
end while
return w

The full iterative algorithm using clustering and constraint generation is shown in Algorithm 5.2. The matrices Q and L in the mathematical program can be pre-computed in each iteration using the matrix of strategies and corresponding T, as well as the set of constraints $\mathcal{C}_i$. The computeAttack() function generates an optimal attack by solving (often approximately) the optimization problem $z_i = \arg\min_{x \in \mathcal{C}_i} l_A(x, x_i; w)$.

Note that convergence of this algorithm is only guaranteed if the full feature space is finite (although it need not be fast). When features are continuous, the constraint generation algorithm need not converge at all, although it likely would in most cases reach a point where new iterations result in very small changes to the classifier.

Next, we illustrate this approach in the context of two of the adversarial models described in Chapter 4. As we show below, both can be formulated as mixed-integer linear programs.

The first model we consider, extending Problem (4.7) in Chapter 4, minimizes the adversary's cost $c(x, x_i)$ subject to the constraint that $w^T x \leq 0$, that is, that the adversarial instance is classified as benign. The extension discussed there is to also impose the cost constraint $\mathcal{C}_i = \{x | c(x, x_i) \leq C\}$. In the notation above, this means that the adversary's loss is just $l_A(x, x_i; w) = c(x, x_i)$. The cost budget constraint can be handled directly by the mathematical program described above. The nonlinear constraint (5.8h) now becomes $(1 - 2r_a)w^T T_a \leq 0$.

While this constraint again introduces bilinear terms, these can be linearized as well since r_a are binary. In particular, we can replace it with the following constraints:

$$\forall a : \sum_j w_j T_{ja} \le 2 \sum_j T_{ja} t_{aj}$$
$$\forall a, j : -M r_a \le t_{aj} \le M r_a$$
$$\forall a, j : w_j - M(1 - r_a) \le t_{aj} \le w_j + M(1 - r_a),$$

(5.9)

where we introduce a new variable t_{aj} to assist in linearization. The full mathematical program for adversarial empirical risk minimization in a sparse SVM thus becomes a mixed-integer linear program (MILP) in the context of this attack model. Finally, we can implement the iterative constraint generation approach by executing a variant of the Lowd and Meek algorithm (Algorithm 4.1) in each iteration in response to the classifier w computed in the prior iteration.

Our second example is the evasion attack model described by Equation (4.6). In this case, adversarial loss becomes $l_A(x, x_i; w) = w^T x + c(x, x_i)$. There is no constraint C_i, and $h(x; w) \equiv 0$, which also eliminates the nonlinear constraint (5.8h). The attacker's best response computation comupteAttack() can be calculated by using a coordinate greedy algorithm (see Chapter 4). Thus, again, we obtain a MILP for computing an optimal learning algorithm for minimizing adversarial empirical risk.

5.2.2 EVASION-ROBUST SVM AGAINST FREE-RANGE ATTACKS

Our next special case of robust learning considers the free-range attack described in the previous chapter. An important characteristic of this attack is that the attacker's goal is to maximize the loss corresponding to adversarial instances in the training data (as a proxy for adversarial risk), subject to a collection of linear constraints. This allows for a tractable extension of the linear SVM optimization problem, where $f(x) = w^T x + b$ for the weight vector w and bias term b (note that we make the bias explicit in this formulation, as the corresponding constant 1 feature would not be subject to an attack).

We start with the *adversarial hinge loss*, defined as follows:

$$h(w, b, x_i) = \begin{cases} \max\limits_{\tilde{x}_i} \max\{0, 1 - (w^T \tilde{x}_i + b)\} & \text{if } y_i = +1 \\ \{0, 1 + (w^T x_i + b)\} & \text{if } y_i = -1 \end{cases}$$
$$\text{s.t.} : \quad C_f x^{min} \le \tilde{x}_i \le C_f x^{max}.$$

(5.10)

Following the standard SVM risk formulation, we then obtain

$$\min_{w,b} \sum_{i \in \mathcal{D}_-} \max \left\{0, 1 + (w^T x_i + b)\right\} + \sum_{i \in \mathcal{D}_+} \max_{\tilde{x}_i} \max \left\{0, 1 - (w^T \tilde{x}_i + b)\right\} + \delta ||w||^2. \quad (5.11)$$

Combining cases for positive and negative instances, this is equivalent to:

$$\min_{w,b} \sum_i \max_{\tilde{x}_i} \max \left\{0, 1 - y_i(w^T x_i + b) - \frac{1}{2}(1 + y_i)w^T(\tilde{x}_i - x_i)\right\} + \delta ||w||^2. \quad (5.12)$$

This is a disjoint bilinear problem with respect to w and $\tilde{x}_i$. Here, we are interested in discovering optimal assignment to $\tilde{x}_i$ for a given w.

The first step is to note that the worst-case hinge loss for a given data point x_i is obtained when $\eta_i = \tilde{x}_i - x_i$ is chosen to minimize its contribution to the margin, which we can formulate as the following linear program:

$$\min_{\eta_i} \frac{1}{2}(1 + y_i)w^T \eta_i$$
$$\text{s.t.} : \quad C_f(x^{min} - x_i) \leq \eta_i \leq C_f(x^{max} - x_i). \quad (5.13)$$

Next, taking the dual of this linear program we obtain a linear program with dual variables u_i and v_i:

$$\min C_f \sum_j \left(v_{ij}(x_j^{max} - x_{ij}) - u_{ij}(x_j^{min} - x_{ij}) \right)$$
$$\text{s.t.} : \quad u_i - v_i = \frac{1}{2}(1 + y_i)w \quad (5.14)$$
$$u_i, v_i \geq 0.$$

This allows us to write the adversarial version of the SVM optimization problem as

$$\underset{w,b,t_i,u_i,v_i}{\text{argmin}} \quad \sum_i \max \left\{ 0, 1 - y_i \cdot (w^T x_i + b) + t_i \right\} + \delta \|w\|^2$$
$$s.t. \quad t_i \geq \sum_j C_f \left(v_{ij}(x_j^{max} - x_{ij}) - u_{ij}(x_j^{min} - x_{ij}) \right) \quad (5.15)$$
$$u_i - v_i = \frac{1}{2}(1 + y_i)w$$
$$u_i, v_i \geq 0.$$

Adding a slack variable and linear constraints to remove the non-differentiality of the hinge loss, we can finally rewrite the problem as the following quadratic program:

$$\underset{w,b,\xi_i,t_i,u_i,v_i}{\text{argmin}} \quad \sum_i \xi_i + \delta \|w\|^2$$
$$s.t. \quad \xi_i \geq 0$$
$$\xi_i \geq 1 - y_i \cdot (w^T x_i + b) + t_i$$
$$t_i \geq \sum_j C_f \left(v_{ij}(x_j^{max} - x_{ij}) - u_{ij}(x_j^{min} - x_{ij}) \right) \quad (5.16)$$
$$u_i - v_i = \frac{1}{2}(1 + y_i)w$$
$$u_i, v_i \geq 0.$$

5.2.3   EVASION-ROBUST SVM AGAINST RESTRAINED ATTACKS

With the restrained attack model, we modify the hinge loss model and solve the problem following the same steps:

$$
h(w, b, x_i) = \begin{cases} \max\limits_{\tilde{x}_i} \max\{0, 1 - (w^T \tilde{x}_i + b)\} & \text{if } y_i = +1 \\ \{0, 1 + (w^T x_i + b)\} & \text{if } y_i = -1 \end{cases}
$$

$$
\text{s.t. :}
$$
$$
|\tilde{x}_i - x_i| \le C_\xi \left(1 - C_\delta \frac{|x_i^t - x_i|}{|x_i| + |x_i^t|}\right) \circ |x_i^t - x_i|
$$
$$
(x_i^t - x_i) \circ (\tilde{x}_i - x_i) \ge 0,
$$

(5.17)

where $\circ$ denotes the pointwise (Hadamard) product.

Again, the worst-case hinge loss is obtained by solving the following minimization problem, where $\eta_i = \tilde{x}_i - x_i$:

$$
\min_{\delta_i} \tfrac{1}{2}(1 + y_i)w^T \eta_i
$$
$$
\text{s.t. :}\quad |\eta_i| \le C_\xi \left(1 - C_\delta \frac{|x_i^t - x_i|}{|x_i| + |x_i^t|}\right) \circ |x_i^t - x_i|
$$
$$
(x_i^t - x_i) \circ \eta_i \ge 0.
$$

(5.18)

If we let

$$
e_{ij} = C_\xi \left(1 - C_\delta \frac{|x_{ij}^t - x_{ij}|}{|x_{ij}| + |x_{ij}^t|}\right)(x_{ij}^t - x_{ij})^2
$$

and multiply the first constraint above by $x_i^t - x_i$ (thereby replacing the nonlinear $|\eta_i|$ absolute value term with a set of equivalent linear inequalities), we obtain the following dual:

$$
\min \sum_j e_{ij} u_{ij}
$$
$$
\text{s.t. :}\quad (-u_i + v_i) \circ (x_i^t - x_i) = \tfrac{1}{2}(1 + y_i)w
$$
$$
u_i, v_i \ge 0.
$$

(5.19)

The SVM risk minimization problem can now be rewritten as follows:

$$
\min_{w, b, t_i, u_i, v_i} \sum_i \max\{0, 1 - y_i(w^T x_i + b) + t_i\} + \delta \|w\|^2
$$
$$
\text{s.t.}\quad t_i \ge \sum_j e_{ij} u_{ij}
$$
$$
(-u_i + v_i) \circ (x_i^t - x_i) = \tfrac{1}{2}(1 + y_i)w
$$
$$
u_i, v_i \ge 0.
$$

(5.20)

Replacing the nonlinear hinge loss with linear constraints, we obtain the following quadratic program:

$$
\begin{aligned}
\min_{w,b,\xi_i,t_i,u_i,v_i} \quad & \sum_i \xi_i + \delta \|w\|^2 \\
s.t. \quad & \xi_i \geq 0 \\
& \xi_i \geq 1 - y_i(w \cdot x_i + b) + t_i \\
& t_i \geq \sum_j e_{ij} u_{ij} \\
& (-u_i + v_i) \circ (x_i^t - x_i) = \tfrac{1}{2}(1 + y_i)w \\
& u_i, v_i \geq 0.
\end{aligned}
\tag{5.21}
$$

5.2.4 EVASION-ROBUST CLASSIFICATION ON UNRESTRICTED FEATURE SPACES

A fairly general alternative approach to evasion-robust classification is offered by Brückner and Scheffer [2011], who make an assumption that the feature space is unrestricted, i.e., $\mathcal{X} = \mathbb{R}^m$. Suppose that the adversary faces the optimization problem of the form

$$
\max_{z_i \in \mathbb{R}^m} \sum_{i \in D_+} l_A(w^T(x_i + z_i)) + \lambda \sum_i Q(z_i),
$$

where $l_A(\cdot)$ and $Q(\cdot)$ are convex with z unconstrained, and the defender's loss function and regularizer are both strictly convex as well.[1] In particular, suppose that

$$
Q(z_i) = \|z_i\|_2^2.
$$

Then the following optimization problem characterizes the Stackelberg equilibrium decisions by the learner and attacker (ignoring regularization by the learner, which can be added without affecting the results):

$$
\sum_{w,\tau} \sum_{i \in D_+} l(w^T(x_i + \tau_i\|w\|^2)) + \sum_{i \in D_-} l(w^T x_i)
$$

$$
\forall i \in D_+ : \tau_i = \frac{2}{\lambda} l(w^T(x_i + \tau_i\|w\|^2)).
$$

If the loss functions are convex and continuously differentiable, this problem can be solved using Sequential Quadratic Programming (SQP).

There are several important limitations of this approach by Bruckner and Schaeffer. The first crucial limitation is the assumption that an adversary can make arbitrary modifications. This is trivially false when the classifier includes a bias feature but, more generally, it is quite

---

[1]The original presentation in Bruckner and Schaefer is slightly more general. Here, we frame it in a way which is consistent with the rest of our discussion of evasion attacks and defenses.

unusual for feature space to be unconstrained; typically, features have upper and lower bounds, and if these are violated, the attack can be easily flagged. The second important limitation is the assumption of strict convexity and continuous differentiability of loss functions, which rules out common loss functions, such as the hinge loss, as well as sparse (l_1) regularization.

5.2.5   ROBUSTNESS TO ADVERSARIALLY MISSING FEATURES

One important special case of attacks is when, at prediction time, an adversarially chosen subset of features is set to zero (or effectively removed). We now describe such an attack, and associated defensive approach, in the context of SVMs. For a given data point (x, y), the attack maximizes hinge loss subject to the constraint that at most K features are removed (set to zero). Let $z_j = 1$ encode the decision to remove jth feature (with $z_j = 0$ otherwise). The attacker's problem can then be captured as the following optimization problem:

$$\max_{z_j \in \{0,1\}} \max\{0, 1 - yw^T(x \circ (1 - z))\}$$
$$\text{s.t.} \sum_j z_j = K. \tag{5.22}$$

Observe that the attacker will always delete the K features with the highest values of $yw_j x_j$. Consequently, we can write the worst-case hinge loss as

$$h^{wc}(yw^T x) = \max\{0, 1 - yw^T x + s\}, \tag{5.23}$$

where

$$s = \max_{z_j \in \{0,1\}, \sum_j z_j = K} yw^T(x \circ z).$$

In addition, since the vertices of the polyhedron defined by $\sum_j z_j = K$ are integral, we can relax the integrality constraint on z_j. Finally, we can change the order of multiplication to obtain

$$s = \max y(w \circ x)^T z$$
$$\text{s.t.}: \quad z \in [0, 1], \sum_j z_j = K. \tag{5.24}$$

Taking the dual of this linear program, we obtain

$$s = \min K z_i + \sum_j v_j \tag{5.25}$$
$$\text{s.t.}: \quad v \geq 0, \quad z_i + v \geq yx \circ w.$$

We can now simply plug this into the standard quadratic program for a linear SVM, with the hinge loss replaced by its worst-case variant $h^{wc}(yw^T x)$, obtaining

$$\min \frac{1}{2}\|w\|_2^2 + C \sum_i \max\{0, 1 - y_i w^T x_i + s_i\}$$
$$s_i \geq Kz_i + \sum_j v_{ij}$$
$$z_i + v_i \geq y_i(w \circ x_i)$$
$$v_i \geq 0.$$

(5.26)

Once the hinge loss term is appropriately linearized, this becomes a convex quadratic program.

5.3   APPROXIMATELY HARDENING CLASSIFIERS AGAINST DECISION-TIME ATTACKS

Even when we are trying to find an optimal evasion-robust *linear* classifier with respect to the specific SVM loss function and l_1 regularization, the problem is exceptionally hard. Even the more tractable approaches for optimal hardening of classifiers against evasion described above make strong assumptions about the attacker, and/or the learning loss function, and are in any case not particularly scalable.

The approach taken in much prior work to address scalability limitations of directly optimizing adversarial empirical risk has aimed at principled approximations. Such techniques fall largely into three potentially overlapping categories:

1. relaxation of the adversarial risk function,

2. relaxation of feature space to continuous features, and

3. iterative retraining with adversarial data.

We first describe relaxation approaches (1 and 2), and then present an iterative retraining approach. Our description of relaxation methods keeps with the binary classification setting to simplify discussion.

5.3.1   RELAXATION APPROACHES

A common relaxation of the adversarial risk function is to turn the game into a *zero-sum* encounter, as follows:

$$\sum_{i \in \mathcal{D}_+} l(f(\mathcal{A}(x_i; f)), +1) + \sum_{i \in \mathcal{D}_-} l(f(x_i), -1) \leq$$
$$\sum_{i \in \mathcal{D}_+} \max_{x \in S(x_i)} l(f(x), +1) + \sum_{i \in \mathcal{D}_-} l(f(x_i), -1),$$

(5.27)

where $S(x_i)$ imposes constraints on modification of x_i (see, e.g., Teo et al. [2007]). A common example of such a constraint is an l_p norm bound $\|x - x_i\|_p^p \leq C$ for an exogenously specified budget C of the attacker.

Consider the l_∞ norm constraints on the attacker as an illustration. Suppose that the feature space is continuous, and consider a score-based variation of the loss function, $l(yg(x))$. Letting $x = x_i + z$ for $z : \|z\|_\infty \leq C$, the zero-sum relaxation becomes

$$\sum_{i \in \mathcal{D}_+} \max_{\|z\|_\infty \leq C} l(g(x_i + z)) + \sum_{i \in \mathcal{D}_-} l(-g(x_i)). \tag{5.28}$$

Suppose $g(x) = w^T x + b$. Then $g(x + z) = w^T x + w^T z + b$, and $\max_{\|z\|_\infty \leq C} l(w^T(x_i + z) + b) = l(w^T x_i - C\|w\|_1 + b)$, if we make the natural assumption that the loss function $l(a)$ is monotonically decreasing in a. Thus, adversarial empirical risk becomes

$$\tilde{R}_A(w) = \sum_{i \in \mathcal{D}_+} l(w^T x_i + b - C\|w\|_1) + \sum_{i \in \mathcal{D}_-} l(-w^T x_i - b). \tag{5.29}$$

In the case of the hinge loss $l(yw^T x) = \max\{0, 1 - yw^T x\}$ (as in support vector machines), this expression now becomes

$$\tilde{R}_A(w) = \sum_{i \in \mathcal{D}_+} \max\{0, 1 - w^T x_i - b + C\|w\|_1\} + \sum_{i \in \mathcal{D}_-} \max\{0, 1 + w^T x_i + b\}. \tag{5.30}$$

Notice the interesting relationship between adversarial robustness to l_∞-norm attacks and sparse regularization: adversarial robustness essentially implies a penalty on the l_1 norm of the weight vector. In any case, we can solve this problem using linear programming. Moreover, we can further "pull" $C\|w\|_1$ out of the loss function, obtaining a standard l_1 regularized linear SVM.

When features are binary, we can modify this approach as follows. Observe that

$$\max_{\|z\|_\infty \leq C} \max\{0, 1 - w^T x_i - b + w^T z\} \leq \max\{1, 1 - w^T x_i - b\} + \max_{\|z\|_\infty \leq C} w^T z. \tag{5.31}$$

Thus, in the case of SVM,

$$\tilde{R}_A(w) \leq \sum_{i \in \mathcal{D}_+} \max\{0, 1 - w^T x_i - b\} + \sum_{i \in \mathcal{D}_- : y_i = -1} \max\{0, 1 + w^T x_i + b\}$$
$$+ |\mathcal{D}_+| \max_{\|z\|_\infty \leq C} w^T z. \tag{5.32}$$

We can further relax

$$\max_{\|z\|_\infty \leq C} w^T z \leq C\|w\|_1 \tag{5.33}$$

(applying the continuous feature relaxation 2 above) so that the attack, again, turns into a corresponding regularization of the model, and, again, we obtain a standard l_1 regularized linear SVM.

These examples illustrate a very important and general connection between regularization and evasion robustness [Russu et al., 2016, Xu et al., 2009b]. To formalize a general connection consider an alternative relaxation of expression (5.27) into a *robust optimization problem*:

$$\sum_{i \in \mathcal{D}_+} l(f(\mathcal{A}(x_i; f)), +1) + \sum_{i \in \mathcal{D}_-} l(f(x_i), -1) \leq \sum_{i \in \mathcal{D}} \max_{x \in S(x_i)} l(f(x), y_i). \tag{5.34}$$

That is, now we allow *every* instance in the domain to be a potential attacker. Further, suppose again that the constraint set $S(x_i) = \{z | \|z - x_i\|_p \leq C\}$ as above. In this case, Xu et al. [2009b] present the following result.

Theorem 5.2 *The following optimization problems are equivalent:*

$$\min_{w,b} \sum_{i \in \mathcal{D}} \max_{z | \|z - x_i\|_p \leq C} \max\{0, 1 - y_i(w^T z + b)\}$$

and

$$\min_{w,b} \sum_{i \in \mathcal{D}} \max\{0, 1 - y_i(w^T x_i + b)\} + C \|w\|_q,$$

where p and q are dual norms (i.e., $\frac{1}{p} + \frac{1}{q} = 1$).

This connection between robustness and regularization is quite powerful. However, the precise connection articulated in Theorem 5.2 is specific to SVMs and the robust optimization problem in which all instances can be adversarial. In practice, the robust optimization formulation itself can be too conservative. Let's return for the moment to our example earlier (Example 5.1). First, suppose that $C = 0.25$. In this case, the optimal threshold $r = 0.5$ is the maximum margin decision rule precisely as we expect from Theorem 5.2 in the one-dimensional special case (where all l_p norms coincide). On the other hand, if $C = 0.4$, robust optimization relaxation of the problem would predict that for any $0.25 \leq r \leq 0.75$ we must make 1 error (since at least one of the two data points will be able to jump across the treshold within the cost budget), and consequently every such threshold is equally good. Of course, this is not the case in the example, since the benign data point will not adversarially change its feature, and, indeed, we can get 0 error by setting $r = 0.3$ (for example; the optimum is not unique).

5.3.2 GENERAL-PURPOSE DEFENSE: ITERATIVE RETRAINING

There is a very general approach for making supervised learning robust to adversarial evasion that has existed in various forms for many years: retraining. The term "retraining" actually refers to a number of different ideas, so henceforth we'll call the specific algorithm below *iterative retraining*. The idea is this: start by applying the standard learning algorithm of choice to training data, followed by transforming each adversarial feature vector in the training data (i.e., each $i : (x_i, y_i) \in S$) according to $\mathcal{A}(x_i; f)$, with the resulting transformed feature vector then *added* to the data, and then iteratively repeating these two steps, either until convergence (or near-convergence), or for a fixed number of iterations. Because of the considerable usefulness of this very simple idea, we present the full iterative retraining algorithm (Algorithm 5.3).

Algorithm 5.3 Iterative Retraining

1: **Input**: training data $\mathcal{D}$; evasion attack function $\mathcal{A}(x; f)$
2: $N_i \leftarrow \emptyset \; \forall \, i : y_i \in T_A$
3: **repeat**
4:     $f \leftarrow \text{Train}(\mathcal{D} \cup_i N_i)$
5:     $v \leftarrow \emptyset$
6:     **for** $i : (x_i, y_i) \in S$ **do**
7:       $x' = \mathcal{A}(x_i; f)$
8:       **if** $x' \notin N_i$ **then**
9:         $v \leftarrow v \cup x'$
10:      **end if**
11:       $N_i \leftarrow N_i \cup x'$
12:     **end for**
13: **until** TerminationCondition(v)
14: **Output**: Learned model f

An important part of Algorithm 5.3 is the TerminationCondition function. One natural termination condition is that $v = \emptyset$, that is, in a given iteration, no new adversarial instances are added to the data. It turns out that if the algorithm actually reaches this condition, the result has the nice theoretical property that the empirical risk of the solution f is an upper bound on optimal adversarial empirical risk.

Theorem 5.3 *Suppose that Algorithm 5.3 has converged where no new adversarial instances can be added, and let $\mathcal{R}_{retraining}$ be the empirical risk after the last iteration. Then $\tilde{\mathcal{R}}_A^* \leq \mathcal{R}_{retraining}$.*

An important thing to notice about the iterative retraining approach is that it does not assume anything about adversarial behavior or learning algorithm. In particular, this idea applies in multi-class classification as well as regression settings. The caveat, of course, is that the algorithm need not always converge, or may converge after a very large number of iterations, effectively making the upper bound in Theorem 5.3 very loose. Nevertheless, considerable empirical experience (e.g., Goodfellow et al. [2015], Li and Vorobeychik [2018]) suggests that this simple idea can be quite effective in practice.

5.4   EVASION-ROBUSTNESS THROUGH FEATURE-LEVEL PROTECTION

An alternative way to induce robustness to evasion is to select a subset of features which are "protected" (e.g., through redundancy or verification), so as to make these impossible for an attacker to modify. This is clearly not universally useful, but is salient in many settings, such as when features correspond to sensor measurements and adversarial attacks involve modifications of these measurements (rather than actual behavior).

We can formalize this problem in the case of binary classification as follows:

$$\min_{f} \min_{r \in \{0,1\}^n : \|r\|_1 \le B} \sum_{i \in \mathcal{D}_+} \max_{x \in S(x_i, r)} l(f(x), +1) + \sum_{i \in \mathcal{D} : y_i = -1} l(f(x_i), -1),$$

where $S(x_i, r)$, the feasible set of attacks, now depends on the choice r of the features which are "protected."

Considering a linear classifier again along with a $\|z\|_\infty \le C$ constraint on the adversary, we can relax this problem to obtain

$$\min_{w} \min_{r \in \{0,1\}^n : \|r\|_1 \le B} \sum_{i \in \mathcal{D}_+} \max_{\|z\|_\infty \le C} l(w^T(x_i + z)) + \sum_{i \in \mathcal{D} : y_i = -1} l(-w^T x_i)$$

$$\le \min_{w, r} \sum_{i \in \mathcal{D}} l(w^T x_i) + |\mathcal{D}_+| C \sum_{j} |w| r_j.$$

We can obtain an approximate solution to this by first computing an optimal w of the problem with $r_j = 0$ for all j, and then greedily selecting $r_j = 1$ for B features with largest $|w_j|$.

5.5   DECISION RANDOMIZATION

The vast majority of techniques for making machine learning robust to decision-time attacks involve modifications of the training procedure to arrive at a more robust fixed model $f \in \mathcal{F}$. Viewing the problem of hardening learning against such attacks through the lens of Stackelberg games, as we had done earlier, suggests another powerful tool, randomization, which has been extensively leveraged by Stackelberg game models in physical security applications [Tambe, 2011]. In this section, we present a principled randomization scheme for supervised binary classification learning in the face of evasion attacks, based on the approach by Li and Vorobeychik [2015].

5.5.1   MODEL

As in the conventional Stackelberg game that represents the adversarial classification problem, the defender acts first, choosing the *probability of flagging* an instance as malicious, followed by attackers (corresponding to malicious instances in the data) choosing optimal evasion strategies.

We denote the defender's decision by a function $q(x)$ which represents the probability of "acting" on a feature vector x (e.g., labeling it as malicious). In the rest of this section, we assume that feature vectors x are binary.

The key idea behind the approach we now describe is to separate the problem of *prediction* from that of *decisions based on predictions*. For the former, we can apply conventional machine learning to the training data $\mathcal{D}$ to obtain a *probabilistic model*, $p(x)$. The semantics of this model is the probability that x is an adversarial "ideal" feature vector *given the current distribution of malicious and benign behavior*. In other words, this represents our belief about the attacker's *preferences*, and is only the first step for the defender. The second step is to identify an optimal *strategy* about which instances x to actually label as malicious, which we allow to be stochastic. This strategy would now take into account adversarial evasion.

To move forward, we now need an evasion attack model. While we discussed a number of candidates in Chapter 4, none are adequate here because they all assume a deterministic classifier f. We now describe a natural adaptation of many of these in the setting with a stochastic decision function for the defender.

Specifically, if the attacker with a preference for x (i.e., for whom $x = x^A$ is the ideal feature vector) chooses an alternative attack vector x', we model the associated utility from successfully bypassing defenses as $V(x)Q(x, x')$, where $Q(x, x') = e^{-\delta||x-x'||}$, with $||\cdot||$ a norm (we use Hamming distance), $V(x)$ the value of the attack, and δ the importance of being close to the preferred x. The full utility function of an attacker with preference x for choosing another input x' when the defense strategy is q is then

$$\mu(x, x'; q) = V(x)Q(x, x')(1 - q(x')), \tag{5.35}$$

since $1 - q(\cdot)$ is the probability that the attacker successfully bypasses the defensive action. The attacker then solves the following optimization problem:

$$v(x; q) = \max_{x'} \mu(x, x'; q). \tag{5.36}$$

Now that we have described the decision model for the attacker, we can turn to the problem that the defender is solving. A natural goal for the defender is to maximize expected value of benign traffic that is classified as benign, less the expected losses due to attacks that successfully bypass the operator.

To formalize, we define the defender's utility function $U_\mathcal{D}(q, p)$ as follows:

$$U_\mathcal{D}(q, p) = \mathbb{E}_x \left[(1 - q(x))(1 - p(x)) - p(x)v(x; q) \right]. \tag{5.37}$$

To interpret the defender's utility function, let us rewrite it for a special case when $V(x) = 1$ and $\delta = \infty$ (so that the attacker will always use the original feature vector x), reducing the utility function to $\mathbb{E}_x[(1 - q(x))(1 - p(x)) - p(x)(1 - q(x))]$. Since $p(x)$ is constant, this is equivalent to minimizing

$$\mathbb{E}_x[q(x)(1 - p(x)) + p(x)(1 - q(x))],$$

which is just the *expected misclassification error*.

5.5.2   OPTIMAL RANDOMIZED OPERATIONAL USE OF CLASSIFICATION

Given the Stackelberg game model of strategic interactions between a defender armed with a classifier, and an attacker attempting to evade it we can now describe an algorithmic approach for solving it. First, approximate the expected utility U_D by taking the sample average utility using the training data; we denote the result by $\tilde{U}_D$. Using $\tilde{U}_D$ as the objective, we can maximize it using the following linear program (LP):

$$\max_q \quad \tilde{U}_D(q, p) \tag{5.38a}$$

$$\text{s.t.} : \quad 0 \leq q(x) \leq 1 \qquad\qquad \forall\, x \in \mathcal{X} \tag{5.38b}$$

$$v(x; q) \geq \mu(x, x'; q) \qquad\qquad \forall\, x, x' \in \mathcal{X}, \tag{5.38c}$$

where constraint (5.38c) computes the attacker's best response (optimal evasion of q), and $\mathcal{X}$ is the full binary feature space.

It is readily apparent that the linear program (5.38) is not a practical solution approach for two reasons: (a) $q(x)$ must be defined over the entire feature space $\mathcal{X}$; and (b) the set of constraints is quadratic in $|\mathcal{X}|$. Since with n features $|\mathcal{X}| = 2^n$, this LP is a non-starter.

The first step toward addressing the scalability issue is to represent $q(x)$ using a set of normalized basis functions, $\{\phi_j(x)\}$, where $q(x) = \sum_j \alpha_j \phi_j(x)$. This allows us to focus on optimizing α_j, a potentially tractable proposition if the set of basis functions is small. With this representation, the LP now takes the following form:

$$\min_{\alpha \geq 0} \sum_j \alpha_j \mathbb{E}[\phi_j(x)(1 - p(x))] + \mathbb{E}[V(x)p(x)Q(x, \alpha)] \tag{5.39a}$$

$$\text{s.t} : Q(x, \alpha) \geq e^{-\delta \|x - x'\|}(1 - \sum_i \alpha_j \phi_j(x')) \quad \forall x, x' \in \mathcal{X} \tag{5.39b}$$

$$\sum_j \alpha_j \leq 1. \tag{5.39c}$$

While we can reduce the number of variables in the optimization problem using a basis representation ϕ, we still retain the intractably large set of inequalities which compute the attacker's best response. To address this issue, suppose that we have an oracle $\mathcal{A}(x; q)$ which can compute a best response x' to a strategy q for an attacker with an ideal attack x. Armed with this oracle, we can use a constraint generation aproach to iteratively compute an (approximately) optimal operational decision function q (Algorithm 5.4 below). The input to Algorithm 5.4 is the feature matrix X in the training data, with X_{bad} denoting this feature matrix restricted to

---

Algorithm 5.4 OptimalRandomizedClassification(X)

ϕ =ConstructBasis()
$\bar{\mathcal{X}} \leftarrow X$
$q \leftarrow$ MASTER($\bar{\mathcal{X}}$)
while true do
  for $x \in X_{bad}$ **do**
    $x' = \mathcal{A}(x; q)$
    $\bar{\mathcal{X}} \leftarrow \bar{\mathcal{X}} \cup x'$
  end for
  if All $x' \in \bar{\mathcal{X}}$ **then**
    return q
  end if
  $q \leftarrow$ MASTER($\bar{\mathcal{X}}$)
end while

"bad" (malicious) instances. At the core of this algorithm is the MASTER linear program which computes an attacker's (approximate) best response using the modified LP (5.39), but using only a small subset of all feature vectors as alternative attacks, which we denote by $\bar{\mathcal{X}}$. The algorithm begins with $\bar{\mathcal{X}}$ initialized to only include feature vectors in the training data X. The first step is to compute an optimal solution, q, with adversarial evasion restricted to X. Then, iteratively, we compute an attacker's best response x' to the current solution q for each malicious instance $x \in X_{bad}$, adding it to $\bar{\mathcal{X}}$, rerun the MASTER linear program to recompute q, and repeat. The process is terminated when we cannot generate any new constraints (i.e., the available constraints already include best responses for the attacker for all malicious instances in training data).

The approach described so far in principle addresses the scalability issues, but leaves two key questions unanswered: (1) how do we construct the basis ϕ, a problem which is of critical importance to good quality approximation (the ConstructBasis() function in Algorithm 5.4); and (2) how do we compute the attacker's best response to q, represented above by an oracle $\mathcal{A}(x, q)$. We discuss these next.

Basis Construction The main idea for the basis representation relies on harmonic (Fourier) analysis of Boolean functions [Kahn et al., 1988, O'Donnell, 2008]. In particular, it is known that every Boolean function $f : \{0, 1\}^n \to \mathbb{R}$ can be uniquely represented as $f(x) = \sum_{S \in B_S} \widehat{f}_S \chi_S(x)$, where $\chi_S(x) = (-1)^{S^T x}$ is a parity function on a given basis $S \in \{0, 1\}^n$, B_S is the set containing all the bases S, and the corresponding Fourier coefficients can be computed as $\widehat{f}_S = \mathbb{E}_x[f(x)\chi_S(x)]$ [De Wolf, 2008, O'Donnell, 2008]. The goal is to approximate $q(x)$ using a Fourier basis. The core task is to compute a set of basis functions to be subsequently used in optimizing $q(x)$. The first step proposed by Li and Vorobeychik [2015] is to uniformly randomly select K feature vectors x_k, use a traditional learning algorithm to obtain the $p(x)$

vector over these, and solve the linear program (5.38) to compute $q(x)$ restricted to these feature vectors. At this point, the same set of feature vectors can be used to approximate a Fourier coefficient of this $q(x)$ for an arbitrary basis S as $t = \frac{1}{m} \sum_{i=1}^{m} q(x^i) \chi_S(x^i)$. We can use this expression to compute a basis set S with the largest and smallest (largest in absolute value) Fourier coefficients using the following integer linear program (replacing the max with the min to get the smallest coefficient):

$$\max_{S} \quad \frac{1}{K} \sum_{k=1}^{K} q(x^k) r_S^k \tag{5.40a}$$

$$s.t.: \quad S^T x^k = 2y^k + h^k \tag{5.40b}$$

$$r_S^k = 1 - 2h^k \tag{5.40c}$$

$$y^k \in Z, h^k \in \{0, 1\}, S \in \{0, 1\}^n. \tag{5.40d}$$

The final basis generation algorithm solves the integer linear program (5.40) for both largest and smallest coefficients iteratively, each time adding a constraint that rules out a previously generated basis, until the absolute value of the optimal solution is sufficiently small.

Computing Adversary's Best Response   The constraint generation algorithm described above presumes the existence of an oracle $\mathcal{A}(x; q)$ which computes (or approximates) an optimal evasion of q (we call this a *best response* to q) for an attacker that would prefer to use a feature vector x. Note that since $V(x)$ is fixed in the attacker's evasion problem (because x is fixed), it can be ignored.

While Li and Vorobeychik [2015] showed that the adversarial evasion problem is strongly NP-Hard, they proposed an effective Greedy heuristic for solving it. This greedy heuristic starts with x and iteratively flips features one at a time, flipping a feature that yields the greatest decrease in $q(x')$ each time.

5.6   EVASION-ROBUST REGRESSION

We close this chapter by discussing an approach for developing evasion-robust regression, based on the decision-time attack on regression discussed in Chapter 4 due to Grosshans et al. [2013]. Specifically, recall from Chapter 4 (Section 4.3.5) that the optimal attack on a linear regression with parameter vector w is

$$\bar{X}^*(w) = X - (\lambda + \|w\|_2^2)^{-1}(Xw - z)w^T.$$

If we were to consider a Stackelberg equilbrium strategy for the learner, we can embed this into the defender's adversarial empirical risk minimization problem as follows:

$$\min_{w}(\bar{X}^*(w)w - y)^T(\bar{X}^*(w)w - y) + \gamma \|w\|_2^2,$$

where we assume that the learner uses both l_2 loss and l_2 regularization. This problem is not necessarily convex, but is smooth and can be approximately solved using nonlinear programming solvers.

5.7   BIBLIOGRAPHIC NOTES

As with evasion attack modeling, Dalvi et al. [2004] present the first significant advance in modeling evasion attack and defense as a game, and offer an approach for making binary classification more robust to evasion. As in many other examples of evasion-robust learning, their evaluation was focused on spam email filtering. Both their game model, and solution approach, are quite distinct from the Stackelberg game model we use. In particular, their algorithm first computes an optimal attack to a conventionally learned model (a standard Naive Bayes classifier), and then computes an optimal defense in response to this evasion attack. The approach, thus, is akin to the first two iterations of asynchronous best response dynamics in games [Fudenberg and Levine, 1998], rather than computation of either a Stackelberg or a Nash equilibrium.

Teo et al. [2007] present what appears to be the first general approach for robust multiclass SVM in the context of decision-time attacks. They model robustness as ensuring prediction invariance to a set of possible manipulations (essentially, transforming the problem into robust optimization), and extend the optimization problem for computing optimal support vector machine structured classifiers in this setting. The resulting quadratic program can be large if the space of possible manipulations is large, but can be solved, in principle, using column generation techniques. Our discussion reduces their approach to the special case of binary classification (which considerably simplifies it). We also frame the Teo et al. [2007] approach as using an upper bound approximation on the optimal classifier hardening problem based on the zero-sum (worst-case loss) relaxation; this is unlike the original paper, which focuses directly on the worst-case loss.

Brückner and Scheffer [2012], in their first foray into adversarial classification, model the problem as a static game between the defender (learner) and attacker, rather than the Stackelberg game which frames all of our discussion. However, in another effort, Brückner and Scheffer [2011] consider the Stackelberg game model; this is the approach we actually describe in this chapter.

The adversarial retraining idea has appeared a number of times in prior literature, in varying incarnations. Its iterative version is systematically described and analyzed by Li and Vorobeychik [2018], and is also discussed by Kantchelian et al. [2016], Grosse et al. [2017], and, in passing, by Teo et al. [2007].

Our discussion of defense against free-range and retrained attacks follows Zhou et al. [2012], who also introduced the associated evasion attack models we described in Chapter 4. Li and Vorobeychik [2014, 2018] introduce both the feature cross-substitution attack discussed in Chapter 4, and the mixed-integer linear programming approach for minimizing adversarial empirical risk in the context of l_1 regularized SVM. Kantarcioglu et al. [2011] discusses how to apply Stackelberg game framework for feature subset selection (i.e., how to choose subset of features that are resistant to adversarial attacks while providing good classification accuracy). Ensemble classifier learning resistant against attacks from multiple adversaries that have different evasion capabilities is discussed in Zhou and Kantarcioglu [2016]. In this work, different classifiers that are optimal against different types of attackers are combined using a Stackelberg game framework.

The connection between robust learning (in the minimax sense we describe here) was first shown by Xu et al. [2009b], with follow-up approaches including Russu et al. [2016] and Demontis et al. [2017a]; the latter used this connection to develop a robust Android malware detector.

The idea of hardening learning against decision-time attacks by protecting specific observed features is a special case of the approach by Alfeld et al. [2017] who consider a general class of defensive actions which restrict the space of manipulations of the observed feature vector.

Our discussion of decision randomization in adversarial learning follows closely the approach by Li and Vorobeychik [2015], who are in turn inspired by the extensive literature on randomization in Stackelberg models of security [Tambe, 2011].

Finally, our description of hardening linear regression against decision-time attacks is due to Grosshans et al. [2013], who consider the more general problem in which the learner is uncertain about cost parameters of the attacker's model which represent the relative importance of different data points to the attacker. We omit discussion of this more complex Bayesian framework to significantly streamline presentation of their approach.

CHAPTER 6

Data Poisoning Attacks

Previously, we studied one broad class of attacks we term *decision-time* attacks, or attacks on machine learning models. A crucial feature of such attacks is that they take place *after* learning, when the learned model is in operational use. We now turn to another broad class of attacks which target the learning *algorithms* by tampering directly with data used for training these.

It is useful to consider several categories of poisoning attacks. We now define four such categories which make important distinctions between adversary's capabilities (what, precisely, the adversary can modify about training data), and attack timing. In particular, while poisoning attack models will typically either impose a constraint on the number of modifications or a modification penalty, they may also constrain what can be modified about the data (e.g., feature vectors and labels, only feature vectors, or only labels), and what kinds of modifications are admissible (e.g., insertion only, or arbitrary modification).

- **Label modification attacks**: these attacks allow the adversary to modify solely the *labels* in supervised learning datasets, but for arbitrary data points, typically subject to a constraint on the total modification cost (e.g., an upper bound on the number of labels that can be changed). The common form of this attack is specific to binary classifiers, and is usually known as a *label flipping attack*.

- **Poison insertion attacks**: in this case, the attacker can add a limited number of arbitrary poisoned feature vectors, with a label they may or may not control (depending on the specific threat model). In unsupervised learning settings, of course, labels do not exist, and the adversary may only contaminate the feature vectors.

- **Data modification attacks**: in these attacks the attacker can modify feature vectors and/or labels for an arbitrary subset of the training data.

- **Boiling frog attacks**: in these attacks, the defender is assumed to iteratively retrain a model. Retraining, in turn, presents an opportunity for the attacker to stealthily guide the model astray over time by injecting a small amount of poison each time so that it makes minimal impact in a particular retraining iteration, but the incremental impact of such attacks over time is significant. Boiling frog attacks can be applied in both supervised and unsupervised settings, although they have typically been studied in the context of unsupervised learning problems.

In this chapter, we first discuss poisoning attacks restricted to binary classifiers which allow us to illustrate two categories of the attacks above: label-flipping attacks and insertion of poisoned data, both specific to support vector machines. Next, we describe poisoning attacks on three unsupervised methods: clustering, PCA, and matrix completion. The penultimate section of this chapter connects data poisoning attacks to machine teaching, and describes a very general framework for such attacks. Finally, we close the chapter with a discussion of black-box poisoning attacks.

6.1   MODELING POISONING ATTACKS

Generically, a poisoning attack begins with a pristine training dataset, which we denote by $\mathcal{D}_0$, and transforms it into another, $\mathcal{D}$. The learning algorithm is then trained on $\mathcal{D}$. Just as in decision-time attacks, the attacker may have two kinds of goals: targeted attacks, in which they wish to induce target labels or decisions for a collection of feature vectors in a target set of instances S, and reliability attacks, in which they wish to maximize prediction or decision error.

In a way analogous to the decision-time attacks, the attackers aspiring to poison a dataset trade off two considerations: achieving a malicious objective and minimizing a measure of modification cost. We will capture the former with a generic attacker risk function $R_A(\mathcal{D}, S)$, which is often going to depend on the learning parameters w obtained by training the model on the poisoned training data $\mathcal{D}$, and where dependence on S will typically be omitted, as it is clear from context. The cost function, in turn, will be denoted by $c(\mathcal{D}_0, \mathcal{D})$.

The attacker's optimization problem typically takes one of the following two forms:

$$\min_{\mathcal{D}} R_A(\mathcal{D}, S) + \lambda c(\mathcal{D}_0, \mathcal{D}) \tag{6.1}$$

or

$$\min_{\mathcal{D}} R_A(\mathcal{D}, S)$$
$$\text{s.t.} : \quad c(\mathcal{D}_0, \mathcal{D}) \leq C \tag{6.2}$$

for some exogenously specified modification cost budget C. Sometimes it will be more convenient to deal with the attacker's *utility* (which they try to maximize) rather than risk function (which is minimized). For this purpose, we define the attacker's utility as $U_A(\mathcal{D}, S) = -R_A(\mathcal{D}, S)$.

Consider a simple illustration of data poisoning in Figure 6.1. The three blue circles comprise "pristine" data on which a true model (lower line in black) can be learned. An attacker can poison this dataset by adding a new datapoint, the red circle, which results in a new poisoned model (dashed line in red) which is quite different from the true model.

6.2   POISONING ATTACKS ON BINARY CLASSIFICATION

Some of the most mature literature on poisoning attacks concerns binary classification problems. As this literature also provides many of the foundations of other poisoning attacks, the binary classification setting offers a natural starting point.

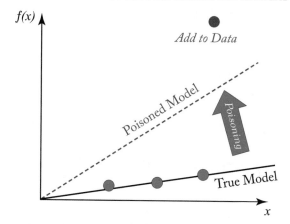

Figure 6.1: Illustration of poisoned linear regression model.

6.2.1 LABEL-FLIPPING ATTACKS

One of the most basic data poisoning attacks one can consider is to change labels for a subset of datapoints in training data. The attacker's goal in this attack is typically to maximize error on unadulterated training data (i.e., data prior to modification); in our terminology, this is a reliability attack. A common motivation behind label-flipping attacks is that datasets used for security may be labeled externally (for example, one can use crowdsourcing to obtain labels for phishing email data), and an attack can thus only pollute the collected labels, but not the feature vectors.

Let $\mathcal{D}_0 = \{(x_i, y_i)\}$ be the original "pristine" training data set. Suppose the attacker has a label-flipping budget C, and the cost of flipping the label of a datapoint i is c_i. Let $z_i = 1$ denote the decision to flip the label of datapoint i, with $z_i = 0$ the decision not to flip this label. Thus, the attacker's modification cost budget can be expressed as

$$c(\mathcal{D}_0, \mathcal{D}) = c(z) = \sum_i z_i c_i \leq C. \tag{6.3}$$

Let $\mathcal{D} = \mathcal{D}(z)$ be the training dataset after a subset of labels chosen by z are flipped. In the most basic variation of the label flipping attack commonly considered in the literature, the target dataset is simply the original training dataset without malicious modification, that is, $S = \mathcal{D}_0$. Ignoring regularization, which can be handled through directly extending the expressions below, the attacker's optimization problem is then

$$\max_z U_A(\mathcal{D}(z)) \equiv \sum_{i \in \mathcal{D}} l(y_i f(x_i; \mathcal{D}(z)))$$

s.t. :

$$f(\mathcal{D}(z)) \in \arg\max_{f'} \sum_{(x_i, y_i) \in \mathcal{D}(z)} l(y_i f(x_i)') \qquad (6.4)$$

$$\sum_i c_i z_i \leq C \quad z_i \in \{0, 1\}.$$

Following Xiao et al. [2012], we now approximate this bi-level problem by recasting it as a single-level optimization problem. At the high level, the attacker aims to induce a large loss of the classifier learned on pristine data, while inducing a classifier which fits well the poisoned data. This can be formalized as

$$\min_z \sum_{(x_i, y_i) \in \mathcal{D}(z)} l(y_i f(x_i; \mathcal{D}(z))) - l(y_i f(x_i; \mathcal{D}_0))$$

$$\text{s.t. :} \sum_i c_i z_i \leq C \quad z_i \in \{0, 1\}. \qquad (6.5)$$

Next, we can represent this mathematical program equivalently by considering a new dataset $\mathcal{D}'$ in which each datapoint x_i is replicated, while the corresponding y_i is flipped, so that $y_{i+n} = -y_i$ for all i, and $\mathcal{D}'$ has $2n$ datapoints. Let $q_i \in \{0, 1\}$ represent which datapoint is chosen, so that for all i, exactly one of q_i and q_{i+n} equals 1. The optimization problem can thus be rewritten as

$$\min_q \sum_{(x_i, y_i) \in \mathcal{D}'} q_i [l(y_i f(x_i; \mathcal{D}')) - l(y_i f(x_i; \mathcal{D}_0))]$$

s.t. :

$$q_i + q_{i+n} = 1 \quad \forall i = 1, \ldots, n \qquad (6.6)$$

$$\sum_{i=n+1}^{2n} c_i q_i \leq C \quad q_i \in \{0, 1\}.$$

As an illustration, we now specialize this problem to a label flipping attack on linear support vector machines. First, observe that $f(x_i; \mathcal{D}_0)$, and associated loss for each data point in $\mathcal{D}'$, can be pre-computed. Let η_i be the corresponding (fixed) loss on a datapoint $(x_i, y_i) \in \mathcal{D}'$.

The problem then becomes

$$\min_{q,w,\epsilon} \sum_{(x_i,y_i)\in\mathcal{D}'} q_i[\epsilon_i - \eta_i] + \gamma \|w\|_2^2$$

$$\text{s.t. :}$$

$$\epsilon_i \geq 1 - y_i w^T x_i \quad \epsilon_i \geq 0$$

$$q_i + q_{i+n} = 1 \quad \forall i = 1,\ldots,n \tag{6.7}$$

$$\sum_{i=n+1}^{2n} c_i q_i \leq C \quad q_i \in \{0,1\}.$$

This results in an integer quadratic program. One approach for approximately optimizing it is alternating minimization, where we alternate between two sub-problems in each iteration.

1. Fix q and minimize over w and ϵ. This becomes a standard quadratic program for a linear SVM.

2. Fix w and ϵ, and minimize over q, which is an integer linear program.

6.2.2 POISON INSERTION ATTACK ON KERNEL SVM

While label flipping attacks are a natural starting place to study poisoning of machine learning models, another important class of attacks is when the adversary can insert a collection of data-points corresponding to feature vectors of its choice, but does not control the labels assigned to these. Consider, for example a spammer, who may choose the nature of spam to send, cognizant of the fact that, in the future, this spam may be used to train a classifier to automatically detect spam. To simplify discussion, we suppose that the adversary inserts only a single datapoint into a training dataset; we can generalize the approach we describe next by having the adversary insert multiple datapoints one at a time.

Consider an original (unaltered) training data set $\mathcal{D}_0$ which is then modified by adding an instance (x_c, y_c) in which the adversary can choose the feature vector x_c but not the label y_c, resulting in a new dataset $\mathcal{D}$. Let a target data set on which the adversary wishes to maximize the learner's risk be denoted by S as before. For the sake of simplifying the discussion, suppose that $S = (x_T, y_T)$, that is, a single target datapoint on which the adversary wishes to induce an error. Observe that the training datset on which a model is learned will now become $\mathcal{D}(x_c) = \mathcal{D}_0 \cup (x_c, y_c)$. Moreover, by allowing the attacker to only add a single feature vector (with a given label) to the data, we effectively impose a budget constraint of inserting a single datapoint; thus, no further explicit discussion of the modification cost is necessary here.

Let $f_{x_c}(x)$ denote the function learned on $\mathcal{D}(x_c)$. The adversary's optimization problem can then be formulated as

$$\max_{x_c} U_A(x_c) \equiv l(y_T f_{x_c}(x_T)). \tag{6.8}$$

Just as above, we now illustrate this attack derived specifically for support vector machines, and allow for arbitrary kernels. First, we introduce some new notation. For a datapoint (x_i, y_i) in training data, define $Q_i(x, y) = y_i y K(x_i, x)$ for some Kernel function $K(\cdot, \cdot)$. In particular, for (x_T, y_T), this becomes $Q_{iT} = y_i y_T K(x_i, x_T)$. As shown by Cauwenberghs and Poggio [2001], the SVM loss function can be expressed as $l(y_T f_{x_c}(x_T)) = \max\{0, 1 - y_T f_{x_c}(x_T)\} = \max\{0, -g_T\}$, where

$$g_T = \sum_{i \in \mathcal{D}_0} Q_{iT} z_i(x_c) + Q_{cT}(x_c) z_c(x_c) + y_T b(x_c) - 1, \qquad (6.9)$$

with z_i and b the dual solutions of Kernel SVM (b is also the bias or intercept term, which we represent explicitly here). Then,

$$f_{x_c}(x) = \sum_i z_i(x_c) y_i K(x_i, x) + b(x_c). \qquad (6.10)$$

The approach to solving this problem, taken by Biggio et al. [2012], is to use gradient ascent, deriving gradients based on the characterization of optimal SVM solutions.

The first challenge which arises with a gradient ascent approach is that hinge loss is not everywhere differentiable, and is constant whenever the defender classifies (x_T, y_T) correctly and outside the SVM classification margin. To address this, we can replace the optimization by a lower bound $-g_T$, solving instead the following problem, where we omit the constant term, and simplify notation by using x in place of x_c:

$$\min_x g_T(x) \equiv \sum_{i \in \mathcal{D}_0} Q_{iT} z_i(x) + Q_{cT}(x) z_c(x) + y_T b(x). \qquad (6.11)$$

The corresponding gradient descent (since we are now minimizing) then involves iterative update steps, where in iteration $t + 1$ we update x as follows:

$$x^{t+1} = x^t - \beta_t \nabla g_T(x^t), \qquad (6.12)$$

where β_t is the learning rate. Then, the gradient of g_T with respect to a given component k of x is

$$\frac{\partial g_T}{\partial x_k} = \sum_{i \in \mathcal{D}_0} Q_{iT} \frac{\partial z_i}{\partial x_k} + z_c \frac{\partial Q_{cT}}{\partial x_k} + Q_{cT} \frac{\partial z_c}{\partial x_k} + y_T \frac{\partial b}{\partial x_k}. \qquad (6.13)$$

In order to make further progress, we again appeal to the special structure of the SVM. Specifically, in the optimal solution of SVM, and the associated KKT conditions, the set of training datapoints can be split into three subsets: R (reserve points, for which $z_i = 0$), S (support vectors, for which $0 < z_i < C$, where C is the weight of the loss term relative to the regularization term), and E (error vectors, with $z_i = C$). For each datapoint i in training data, let

$$g_i = \sum_{j \in \mathcal{D}_0} Q_{ij} z_j + y_i b - 1. \tag{6.14}$$

From the SVM KKT conditions, for all $i \in R$, $g_i > 0$, for $i \in S$, $g_i = 0$, and for $i \in R$, $g_i < 0$. Moreover,

$$h = \sum_{j \in \mathcal{D}_0} y_j z_j = 0. \tag{6.15}$$

Now, we attempt to change x so as to preserve solution optimality, which we can do if we ensure that the sets R, S, and E remain the same. If this is the case, then for any $i \in R \cup E$, $\partial z_i / \partial x_k = 0$, since z_i must remain a constant. Consequently, for any $i \in S$,

$$\frac{\partial g_i}{\partial x_k} = \sum_{j \in S} Q_{ij} \frac{\partial z_j}{\partial x_k} + \frac{\partial Q_{ic}}{\partial x_k} z_c + y_i \frac{\partial b}{\partial x_k} = 0 \tag{6.16}$$

and

$$\frac{\partial h}{\partial x_k} = \sum_{j \in \mathcal{D}_0} y_j \frac{\partial z_j}{\partial x_k}. \tag{6.17}$$

Converting to matrix-vector notation, let $\frac{\partial z_S}{\partial x_k}$ be the vector of partial derivatives of $z_j(x)$ with respect to x_k for all $j \in S$, let Q_S be a matrix of Q_{ij} for $i, j \in S$, and let y_S be the vector of y_i for $i \in S$. Finally, let Q_{Sc} be a vector of Q_{ic} for $i \in S$. We can then write these conditions as

$$Q_S \frac{\partial z_S}{\partial x_k} + \frac{\partial Q_{Sc}}{\partial x_k} z_c + y_S \frac{\partial b}{\partial x_k} = 0 \tag{6.18}$$

and

$$y_S^T \frac{\partial z_S}{\partial x_k} = 0, \tag{6.19}$$

and can solve for $\frac{\partial b}{\partial x_k}$ and $\frac{\partial z_S}{\partial x_k}$ as follows:

$$\begin{bmatrix} \frac{\partial b}{\partial x_k} \\ \frac{\partial z_S}{\partial x_k} \end{bmatrix} = -z_c \begin{bmatrix} 0 & y_S^T \\ y_S & Q_S \end{bmatrix}^{-1} \begin{bmatrix} 0 \\ \frac{\partial Q_{Sc}}{\partial x_k} \end{bmatrix}. \tag{6.20}$$

To complete the calculation of the gradient, we need $\partial Q_{ic}/\partial x_k$ and $\partial Q_{Tc}/\partial x_k$, which amounts to taking derivatives of the Kernel function. The full algorithm then proceeds by iterating the following two steps.

1. Learn SVM (perhaps incrementally) using $\mathcal{D}_0 \cup x_t$ (the value of poisoned feature vector x from previous step t) and

2. update $x^{t+1} = x^t - \beta_t \nabla g_T(x^t)$ using the gradient derived above.

6.3   POISONING ATTACKS FOR UNSUPERVISED LEARNING

Unlike supervised learning, unsupervised settings involve a dataset comprised solely of feature vectors $\mathcal{D} = \{x_i\}$. As before, we let the original "pristine" dataset be denoted by $\mathcal{D}_0$, and let $\mathcal{D}$ denote the transformed dataset (which may include new datapoints).

Three problems in adversarial unsupervised learning have received particular attention in the context of poisoning: clustering (for example, when used to cluster malware), anomaly detection, and matrix completion. We consider the first two in this section, and devote the following section to an in-depth discussion of attacks on matrix completion.

6.3.1   POISONING ATTACKS ON CLUSTERING

A clustering algorithm can be generically represented as a mapping $f(\mathcal{D})$ which takes a dataset $\mathcal{D}$ as input and returns a cluster assignment. For many clustering algorithms, we can represent a cluster assignment by a matrix $\mathbf{Y}$ where an entry y_{ik} is the probability that a datapoint i is assigned to a cluster k. In most common clustering algorithms, y_{ik} are binary, indicating the cluster assignments of datapoints. Let us denote the clustering assignment for a pristine dataset $\mathcal{D}_0$ by $\mathbf{Y}_0$. For a poisoned dataset $\mathcal{D}$, let $\mathbf{Y} = f(\mathcal{D})$ be the resulting cluster assignment. To simplify discussion, suppose that the attacker only *modifies* $\mathcal{D}_0$.

As in supervised settings, we can consider two goals for the attacker: targeted and reliability attacks. In a targeted attack, the attacker has a target clustering $\mathbf{Y}_T$, and they wish to get as close to this target as possible. A special case would be a stealthy attack in which specific target data points are clustered incorrectly without altering cluster assignments of other data. In a reliability attack, the attacker wishes to maximally distort the cluster assignment based on original data.

In order to devise meaningful measures of success for an attacker or the learner, we need to account for the fact that the particular cluster identities are entirely arbitrary. In fact, what is non-arbitrary is a *joint* assignment of feature vectors to the *same* cluster. We can capture this by using $\mathbf{O}_0 = \mathbf{Y}_0\mathbf{Y}_0^T$ instead of $\mathbf{Y}_0$ as the outcome measure. Thus, if both i and j are in the same cluster k, i.e., $y_{0ik} = y_{0jk} = 1$, then $[\mathbf{O}_0]_{ij} = 1$. Similarly, we define $\mathbf{O} = \mathbf{Y}\mathbf{Y}^T$ as the outcome (pairwise assignment of players to the same cluster) for the clustering induced by the poisoning attack, and $\mathbf{O}_T = \mathbf{Y}_T\mathbf{Y}_T^T$ to represent the target outcome in a targeted attack.

We can formally model either attack by endowing an attacker with a risk function $R_A(\mathbf{O}_0, \mathbf{O})$ with original and induced assignments as arguments. For a targeted attack, a natural risk function is distance of induced clustering outcomes to the target outcome:

$$R_A(\mathbf{O}_0, \mathbf{O}) = \|\mathbf{O} - \mathbf{O}_T\|_F, \tag{6.21}$$

where $\|\cdot\|_F$ is the Frobenius norm. Similarly, for a reliability attack, the risk function can be the similarity (negative distance) between the correct and induced outcomes:

$$R_A(\mathbf{O}_0, \mathbf{O}) = -\|\mathbf{O} - \mathbf{O}_0\|_F. \tag{6.22}$$

While the attacker aims to change the clustering outcome, they also face costs and/or constraints associated with the attack. We can capture the costs of modifying training data by a cost function:

$$c(\mathcal{D}_0, \mathcal{D}) = c(\mathbf{X}_0, \mathbf{X}) = \|\mathbf{X}_0 - \mathbf{X}\|_F, \tag{6.23}$$

where $\mathbf{X}_0$ and $\mathbf{X}$ are the original and poisoned feature matrices. This gives rise to two alternative formulations of the attack on clustering. The first minimizes a combination of loss and cost:

$$\min_{\mathbf{X}} R_A(\mathbf{O}_0, \mathbf{O}(\mathbf{X})) + \lambda c(\mathbf{X}_0, \mathbf{X}), \tag{6.24}$$

where we make explicit the dependence of the outcome clustering after poisoning on the poisoned dataset $\mathbf{X}$. The second minimizes adversary's risk subject to a cost constraint:

$$\min_{\mathbf{X}} R_A(\mathbf{O}_0, \mathbf{O}(\mathbf{X}))$$
$$\text{s.t.} : c(\mathbf{X}_0, \mathbf{X}) \leq C. \tag{6.25}$$

In general, the poisoning attack on clustering is extremely challenging computationally because clustering itself is a highly non-trivial optimization problem. However, there are several important special cases of such attacks for which effective heuristics have been proposed in prior literature. The first is a reliability attack in which we wish to add a collection of C datapoints to a dataset so as to maximally distort an original cluster assignment, in the context of agglomerative clustering [Biggio et al., 2014a,b]. The idea is to add a single datapoint at a time in a way that bridges a pair of nearby clusters. In particular, for k clusters we can define $k - 1$ bridges from each cluster to its closest neighboring cluster. Taking the corresponding pairs of points, we can define the shortest bridge as the shortest distance between any two points belonging to different clusters. Adding a datapoint midway between these will make it most likely that the corresponding two clusters are merged. Thus, by iteratively adding points, we can significantly distort the original cluster assignment.

The second special case is a targeted attack in which we target a specific collection of datapoints for which we wish to induce an incorrect clustering without affecting cluster assignments of any other datapoints [Biggio et al., 2014b]. Suppose that x_i is a feature vector which the attacker aims to shift to a different cluster, and let d be the point in the target cluster closest to x_i. If our cost budget constraint is that no datapoint can be modified in l_2 norm by more than C, we can transform x_i into $x_i + \gamma(d - x_i)$ where $\gamma = \min(1, C/\|d - x_i\|_2)$.

6.3.2   POISONING ATTACKS ON ANOMALY DETECTION

Attacks on Online Centroid Anomaly Detection   The first attack we discuss is on centroid anomaly detection, where the mean is computed online, and is due to Kloft and Laskov [2012]. Their attack falls into the category of *boiling frog attacks* we mentioned in the beginning of the chapter. Specifically, they assume that the anomaly detector is periodically retrained as new data is collected, and the adversary adds datapoints between each retraining iteration.

In this attack, the attacker has a target feature vector x_T which they wish to use in the future, and would like to ensure that it is misclassified as normal. Formally, the goal is that $\|x_T - \mu_t\| \geq r$ at some learning iteration t, where μ_t is the centroid mean and r the threshold of the centroid anomaly detector (see Chapter 2, Section 2.2.4 for details). In our terminology, this is an *targeted attack*, with the goal of the attacker represented by a target centroid mean μ_T such that $\|x_T - \mu_T\| = r$.

Kloft and Laskov [2012] propose a *greedy-optimal* strategy for incrementally poisoning online centroid-based anomaly detectors. Under this strategy, in each iteration when the attacker is able to insert an attack instance into the training data, they insert an instance along the line connecting the current centroid μ_t with the attack target feature vector x_T, and exactly on the boundary of the normal region. Formally, suppose that the adversary can insert poison in iteration t and define $a = \frac{x_T - \mu_t}{\|x_T - \mu_t\|}$ as the unit direction from the mean μ_t to x_T. The greedy-optimal attack in this iteration is then $x' = \mu_t + ra$, which maximally displaces the centroid toward the attack target, but remains in the normal region to ensure that it is not simply discarded by the current anomaly detector.

PCA   Next, we describe an example attack on PCA-based anomaly detectors (see Chapter 2, Section 2.2.4 for details) based on Rubinstein et al. [2009]. In this setting, the attacker aims to execute a denial-of-service (DoS) attack, which implies adding anomalous-looking traffic. In our representation, this corresponds to adding an amount δz to the original traffic, where δ is the strength of the attacker, and z the feature-level impact. If we assume that the attacker knows corresponding future background traffic x, then the attacker would need to perturb the resulting traffic x into $x' = x + \delta z$ to successfully execute the DoS attack. The attacker's goal is to maximize δ *in this future DoS attack*, by skewing or stretching the anomaly detector so that the future attack appears normal.

In the poisoning attack, suppose that the attacker can modify the content of the original data by adding a matrix $\tilde{\mathbf{X}} \in \mathbb{X}$ into the original dataset $\mathbf{X}_0$, with the constraint that $\|\tilde{\mathbf{X}}\|_1 \leq C$, where C is the attacker's budget constraint and $\mathbb{X}$ the set of feasible modifications. Let r be the threshold of the anomaly detector. Then we can represent the attacker's optimization problem as

$$\max_{\delta, \tilde{\mathbf{X}} \in \mathbb{X}} \quad \delta$$
$$\text{s.t.:} \quad \mathbf{V} = \text{PCA}(\mathbf{X} + \tilde{\mathbf{X}})$$
$$\|(\mathbb{I} - \mathbf{V}\mathbf{V}^T)(x + \delta z))\| \leq r \tag{6.26}$$
$$\|\tilde{\mathbf{X}}\|_1 \leq C.$$

While this problem is intractable as is, we can approximate the objective as maximizing the magnitude of the projected direction of the attack, $\|(\mathbf{X}_0 + \tilde{\mathbf{X}})z\|_2^2$, yielding

$$\max_{\tilde{\mathbf{X}} \in \mathbb{X}} \quad \|(\mathbf{X}_0 + \tilde{\mathbf{X}})z\|_2^2$$
$$\text{s.t.:} \quad \|\tilde{\mathbf{X}}\|_1 \leq C. \tag{6.27}$$

This problem, in turn, can be solved using projected gradient ascent (also known as projection pursuit).

6.4   POISONING ATTACK ON MATRIX COMPLETION

6.4.1   ATTACK MODEL

In this section we describe the framework for poisoning matrix completion algorithms, based on Li et al. [2016]. Recall from Chapter 2 that in the matrix completion problem, we begin with an $n \times m$ matrix $\mathbf{M}$ in which rows correspond to n users and columns correspond to m items (with the semantics that each user has an opinion about each item). However, we only observe a small proportion of entries in $\mathbf{M}$, corresponding to actual ratings, and the goal is to infer the rest—that is, to complete the matrix.

In the attack model we now discuss, the attacker is capable of adding $\lfloor \alpha n \rfloor$ malicious users to the training data matrix, and each malicious user is allowed to report its preference on at most C items with each preference bounded in the range $[-\Lambda, \Lambda]$.

Let $\mathbf{M}_0 \in \mathbb{R}^{n \times m}$ denote the original data matrix and $\widetilde{\mathbf{M}} \in \mathbb{R}^{n' \times m}$ to denote the data matrix of all $n' = \alpha n$ malicious users. Let $\widetilde{\Omega}$ be the set of non-zero entries in $\widetilde{\mathbf{M}}$ and $\widetilde{\Omega}_i \subseteq [m]$ be all items that the ith malicious user rated. According to our attack models, $|\widetilde{\Omega}_i| \leq C$ for every $i \in \{1, \cdots, n'\}$ and $\|\widetilde{\mathbf{M}}\|_{\max} = \max |\widetilde{\mathbf{M}}_{ij}| \leq \Lambda$. For an arbitrary matrix $\mathbf{M}$, let Ω be the subset of entries that are observed, and recall from Section 2.2.3 that the notation $R_\Omega(\mathbf{M})$ means that $[R_\Omega(\mathbf{M})]_{ij}$ equals $\mathbf{M}_{ij}$ if $(i, j) \in \Omega$ and 0 otherwise.

Let $\Theta_\gamma(\mathbf{M}_0; \widetilde{\mathbf{M}})$ be the optimal solution computed jointly on the original and poisoned data matrices $(\mathbf{M}_0; \widetilde{\mathbf{M}})$ using regularization parameters γ. For example, Eq. (2.6) becomes

$$\Theta_\gamma(\mathbf{M}_0; \widetilde{\mathbf{M}}) = \underset{\mathbf{U}, \widetilde{\mathbf{U}}, \mathbf{V}}{\arg\min} \|R_\Omega(\mathbf{M}_0 - \mathbf{U}\mathbf{V}^\top)\|_F^2 + \|R_{\widetilde{\Omega}}(\widetilde{\mathbf{M}} - \widetilde{\mathbf{U}}\mathbf{V}^\top)\|_F^2$$
$$+ 2\gamma_U(\|\mathbf{U}\|_F^2 + \|\widetilde{\mathbf{U}}\|_F^2) + 2\gamma_V\|\mathbf{V}\|_F^2, \tag{6.28}$$

where the resulting Θ consists of low-rank latent factors $\mathbf{U}, \widetilde{\mathbf{U}}$ for normal and malicious users as well as $\mathbf{V}$ for items. Simiarly, for the nuclear norm minimization formulation in Eq. (2.7), we have

$$\Theta_\gamma(\mathbf{M}_0; \widetilde{\mathbf{M}}) = \operatorname*{argmin}_{\mathbf{X}, \widetilde{\mathbf{X}}} \|\mathcal{R}_\Omega(\mathbf{M}_0 - \mathbf{X})\|_F^2 + \|\mathcal{R}_{\tilde{\Omega}}(\widetilde{\mathbf{M}} - \widetilde{\mathbf{X}})\|_F^2 + 2\gamma\|(\mathbf{X}; \widetilde{\mathbf{X}})\|_*. \quad (6.29)$$

Here, the solution is $\Theta = (\mathbf{X}, \widetilde{\mathbf{X}})$.

Let $\widehat{\mathbf{M}}(\Theta)$ be the matrix estimated from the learned model Θ. For example, for Eq. (6.28) we have $\widehat{\mathbf{M}}(\Theta) = \mathbf{U}\mathbf{V}^\top$ and for Eq. (6.29) we have $\widehat{\mathbf{M}}(\Theta) = \mathbf{X}$. The goal of the attacker is to find optimal malicious users $\widetilde{\mathbf{M}}^*$ such that

$$\widetilde{\mathbf{M}}^* \in \operatorname*{argmax}_{\widetilde{\mathbf{M}} \in \mathbb{M}} U(\widehat{\mathbf{M}}(\Theta_\gamma(\mathbf{M}_0; \widetilde{\mathbf{M}})), \mathbf{M}_0). \quad (6.30)$$

Here $\mathbb{M} = \{\widetilde{\mathbf{M}} \in \mathbb{R}^{n' \times m} : |\tilde{\Omega}_i| \leq C, \|\widetilde{\mathbf{M}}\|_{\max} \leq \Lambda\}$ is the set of all feasible poisoning attacks discussed earlier in this section and $U(\widehat{\mathbf{M}}, \mathbf{M}_0)$ denotes the attacker's utility for diverting the collaborative filtering algorithm to predict $\widehat{\mathbf{M}}$ on an original data set $\mathbf{M}_0$, with the help of few malicious users $\widetilde{\mathbf{M}}$.

One can consider several objectives for the attacker in the context of poisoning matrix completion. The first is a *reliability attack,* in which the attacker wishes to maximize the error of the collaborative filtering system. To formally define this attack objective, suppose that $\overline{\mathbf{M}}$ is the prediction of the collaborative filtering system without data poisoning attacks.[1] The attacker's risk function is then defined as the total amount of perturbation of predictions between $\overline{\mathbf{M}}$ and $\widehat{\mathbf{M}}$ (predictions after poisoning attacks) on unseen entries Ω^C:

$$U^{\mathrm{rel}}(\mathbf{M}_0, \widehat{\mathbf{M}}) = \|\mathcal{R}_{\Omega^C}(\widehat{\mathbf{M}} - \overline{\mathbf{M}})\|_F^2. \quad (6.31)$$

Another class of objectives we can consider are targeted attacks, in which the attacker wishes to boost or reduce the popularity of a (subset) of items, respectively.[2] To formalize, suppose $S \subseteq [m]$ is the subset of items the attacker is interested in and w is a pre-specified weight vector by the attacker with w_j a weight of item $j \in S$ (positive for items whose rating the attacker wishes to boost, and negative for those whose rating the attacker aims to reduce). The utility function is

$$U_{S,w}^{\mathrm{targeted}}(\widehat{\mathbf{M}}, \mathbf{M}_0) = \sum_{i=1}^{m} \sum_{j \in S} w_j \widehat{\mathbf{M}}_{ij}. \quad (6.32)$$

Finally, we can consider a hybrid attack:

$$U_{S,w,\mu}^{\mathrm{hybrid}}(\mathbf{M}_0, \widehat{\mathbf{M}}) = \mu_1 U^{\mathrm{rel}}(\mathbf{M}_0, \widehat{\mathbf{M}}) + \mu_2 U_{S,w}^{\mathrm{targeted}}(\mathbf{M}_0, \widehat{\mathbf{M}}), \quad (6.33)$$

---

[1]Note that when the collaborative filtering algorithm and its parameters are set, $\overline{\mathbf{M}}$ is a function of observed entries $\mathcal{R}_\Omega(\mathbf{M}_0)$.
[2]We remark here that the notion of targeted attacks we deal here does not quite fit into our definition of targeted attacks, which assume a single target. We keep the simpler notion to simplify discussion, particularly as that accounts for the large majority of targeted attacks, but note that a more general definition would consider target label *sets,* in which any label in the target set is satisfactory for the attacker.

where $\mu = (\mu_1, \mu_2)$ are coefficients that trade off the two attack objectives (reliability and targeted attacks). In addition, μ_1 could be negative, which models the case when the attacker wants to leave a "light trace:" the attacker wants to make his item more popular while making the other recommendations of the system less perturbed to avoid detection.

Next we describe practical algorithms to solve the optimization problem in Eq. (6.30). We first consider the alternating minimization formulation in Eq. (6.28) and derive a projected gradient ascent method that solves for the corresponding optimal attack strategy. Similar derivations are then extended to the nuclear norm minimization formulation in Eq. (6.29). Finally, we discuss how to design malicious users that mimic normal user behavior in order to avoid detection.

6.4.2   ATTACKING ALTERNATING MINIMIZATION

We now describe a *projected gradient ascent* (PGA) method for solving the optimization problem in Eq. (6.30) with respect to the alternating minimization formulation in Eq. (6.28), which is due to Li et al. [2016]. In particular, in iteration t the algorithm updates $\widetilde{\mathbf{M}}^{(t)}$ as follows:

$$\widetilde{\mathbf{M}}^{(t+1)} = \mathrm{Proj}_{\mathbb{M}}\left(\widetilde{\mathbf{M}}^{(t)} + \beta_t \nabla_{\widetilde{\mathbf{M}}} U(\mathbf{M}_0, \widehat{\mathbf{M}})\right), \tag{6.34}$$

where $\mathrm{Proj}_{\mathbb{M}}(\cdot)$ is the projection operator onto the feasible region $\mathbb{M}$ and β_t is the step size in iteration t. Note that the estimated matrix $\widehat{\mathbf{M}}$ depends on the model $\mathbf{\Theta}_\gamma(\mathbf{M}_0; \widetilde{\mathbf{M}})$ learned on the joint data matrix, which further depends on the malicious users $\widetilde{\mathbf{M}}$. Since the constraint set $\mathbb{M}$ is highly non-convex, we can generate C items uniformly at random for each malicious user to rate. The $\mathrm{Proj}_{\mathbb{M}}(\cdot)$ operator then reduces to projecting each malicious users' rating vector onto an ℓ_∞ ball of diameter Λ, which can be easily evaluated by truncating all entries in $\widetilde{\mathbf{M}}$ at the level of $\pm\Lambda$.

We next show how to (approximately) compute $\nabla_{\widetilde{\mathbf{M}}} U(\mathbf{M}_0, \widehat{\mathbf{M}})$. This is challenging because one of the arguments in the loss function involves an implicit optimization problem. We first apply chain rule to arrive at

$$\nabla_{\widetilde{\mathbf{M}}} U(\mathbf{M}_0, \widehat{\mathbf{M}}) = \nabla_{\widetilde{\mathbf{M}}} \mathbf{\Theta}_\gamma(\mathbf{M}_0; \widetilde{\mathbf{M}}) \nabla_{\mathbf{\Theta}} U(\mathbf{M}_0, \widehat{\mathbf{M}}). \tag{6.35}$$

The second gradient (with respect to $\mathbf{\Theta}$) is easy to evaluate, as all loss functions mentioned in the previous section are smooth and differentiable. On the other hand, the first gradient term is much harder to evaluate because $\mathbf{\Theta}_\gamma(\cdot)$ is an optimization procedure. Fortunately, we can exploit the KKT conditions of the optimization problem $\mathbf{\Theta}_\gamma(\cdot)$ to approximately compute

---

Algorithm 6.5 Optimizing $\widetilde{\mathbf{M}}$ via PGA

1: **Input**: Original partially observed $n \times m$ data matrix $\mathbf{M}_0$, algorithm regularization parameter γ, attack budget parameters α, C and Λ, attacker's utility function U, step size $\{\beta_t\}_{t=1}^{\infty}$.

2: **Initialization**: random $\widetilde{\mathbf{M}}^{(0)} \in \mathbb{M}$ with both ratings and rated items uniformly sampled at random; $t = 0$.

3: **while** $\widetilde{\mathbf{M}}^{(t)}$ does not converge **do**

4:     Compute the optimal solution $\mathbf{\Theta}_\gamma(\mathbf{M}_0; \widetilde{\mathbf{M}}^{(t)})$.

5:     Compute gradient $\nabla_{\widetilde{\mathbf{M}}} U(\mathbf{M}_0, \widehat{\mathbf{M}})$ using Eq. (6.34).

6:     Update: $\widetilde{\mathbf{M}}^{(t+1)} = \mathrm{Proj}_{\mathbb{M}}(\widetilde{\mathbf{M}}^{(t)} + \beta_t \nabla_{\widetilde{\mathbf{M}}} U(\mathbf{M}_0, \widehat{\mathbf{M}}))$.

7:     $t \leftarrow t + 1$.

8: **end while**

9: **Output**: $n' \times m$ malicious matrix $\widetilde{\mathbf{M}}^{(t)}$.

$\nabla_{\widetilde{\mathbf{M}}} \mathbf{\Theta}_\gamma(\mathbf{M}_0; \widetilde{\mathbf{M}})$. More specifically, the optimal solution $\mathbf{\Theta} = (\mathbf{U}, \widetilde{\mathbf{U}}, \mathbf{V})$ of Eq. (6.28) satisfies

$$
\begin{aligned}
\gamma_U \boldsymbol{u}_i &= \sum_{j \in \Omega_i} (\mathbf{M}_{0ij} - \boldsymbol{u}_i^\top \boldsymbol{v}_j) \boldsymbol{v}_j; \\
\gamma_U \tilde{\boldsymbol{u}}_i &= \sum_{j \in \widetilde{\Omega}_i} (\widetilde{\mathbf{M}}_{ij} - \tilde{\boldsymbol{u}}_i^\top \boldsymbol{v}_j) \boldsymbol{v}_j; \\
\gamma_V \boldsymbol{v}_j &= \sum_{i \in \Omega'_j} (\mathbf{M}_{0ij} - \boldsymbol{u}_i^\top \boldsymbol{v}_j) \boldsymbol{u}_i + \sum_{i \in \widetilde{\Omega}'_j} (\widetilde{\mathbf{M}}_{ij} - \tilde{\boldsymbol{u}}_i^\top \boldsymbol{v}_j) \tilde{\boldsymbol{u}}_i,
\end{aligned}
\tag{6.36}
$$

where $\boldsymbol{u}_i, \tilde{\boldsymbol{u}}_i$ are the ith rows (of dimension k) in $\mathbf{U}$ or $\widetilde{\mathbf{U}}$ and $\boldsymbol{v}_j$ is the jth row (also of dimension k) in $\mathbf{V}$. Consequently, $\{\boldsymbol{u}_i, \tilde{\boldsymbol{u}}_i, \boldsymbol{v}_j\}$ can be expressed as functions of the original and malicious data matrices $\mathbf{M}_0$ and $\widetilde{\mathbf{M}}$. Using the fact that $(\boldsymbol{a}^\top \boldsymbol{x})\boldsymbol{a} = (\boldsymbol{a}\boldsymbol{a}^\top)\boldsymbol{x}$ and $\mathbf{M}_0$ does not change with $\widetilde{\mathbf{M}}$, we obtain

$$
\begin{aligned}
\frac{\partial \boldsymbol{u}_i(\widetilde{\mathbf{M}})}{\partial \widetilde{\mathbf{M}}_{ij}} &= \mathbf{0}; \quad \frac{\partial \tilde{\boldsymbol{u}}_i(\widetilde{\mathbf{M}})}{\partial \widetilde{\mathbf{M}}_{ij}} = \left(\gamma_U \mathbf{I}_k + \mathbf{\Sigma}_U^{(i)}\right)^{-1} \boldsymbol{v}_j; \\
\frac{\partial \boldsymbol{v}_j(\widetilde{\mathbf{M}})}{\partial \widetilde{\mathbf{M}}_{ij}} &= \left(\gamma_V \mathbf{I}_k + \mathbf{\Sigma}_V^{(j)}\right)^{-1} \boldsymbol{u}_i.
\end{aligned}
\tag{6.37}
$$

Here $\mathbf{\Sigma}_U^{(i)}$ and $\mathbf{\Sigma}_V^{(j)}$ are defined as

$$
\mathbf{\Sigma}_U^{(i)} = \sum_{j \in \Omega_i \cup \widetilde{\Omega}_i} \boldsymbol{v}_j \boldsymbol{v}_j^\top, \quad \mathbf{\Sigma}_V^{(j)} = \sum_{i \in \Omega'_j \cup \widetilde{\Omega}'_j} \boldsymbol{u}_i \boldsymbol{u}_i^\top.
\tag{6.38}
$$

The full optimization algorithm can then be described in Algorithm 6.5.

6.4.3 ATTACKING NUCLEAR NORM MINIMIZATION

We can also extend the projected gradient ascent algorithm described above to compute optimal attack strategies for the nuclear norm minimization formulation in Eq. (6.29). Since the objective in Eq. (6.29) is convex, the globally optimal solution $\boldsymbol{\Theta} = (\mathbf{X}, \widetilde{\mathbf{X}})$ can be obtained by conventional convex optimization procedures such as proximal gradient descent (a.k.a. singular value thresholding [Cai et al., 2010] for nuclear norm minimization). In addition, the resulting estimation $(\mathbf{X}; \widetilde{\mathbf{X}})$ is low rank due to the nuclear norm penalty [Candès and Recht, 2007].

Suppose $(\mathbf{X}; \widetilde{\mathbf{X}})$ has rank $\rho \leq \min(n, m)$. We can use $\boldsymbol{\Theta}' = (\mathbf{U}, \widetilde{\mathbf{U}}, \mathbf{V}, \boldsymbol{\Sigma})$ as an alternative characterization of the learned model with a reduced number of parameters. Here $\mathbf{X} = \mathbf{u}\boldsymbol{\Sigma}\mathbf{V}^{\top}$ and $\widetilde{\mathbf{X}} = \widetilde{\mathbf{U}}\boldsymbol{\Sigma}\mathbf{V}^{\top}$ are singular value decompositions of $\mathbf{X}$ and $\widetilde{\mathbf{X}}$; that is, $\mathbf{U} \in \mathbb{R}^{n \times \rho}$, $\widetilde{\mathbf{U}} \in \mathbb{R}^{m' \times \rho}$, $\mathbf{V} \in \mathbb{R}^{m \times \rho}$ have orthornormal columns and $\boldsymbol{\Sigma} = \mathbf{diag}(\sigma_1, \cdots, \sigma_\rho)$ is a non-negative diagonal matrix.

To compute the gradient $\nabla_{\widetilde{\mathbf{M}}} U(\mathbf{M}_0, \widehat{\mathbf{M}})$, we again apply the chain rule to decompose the gradient into two parts:

$$\nabla_{\widetilde{\mathbf{M}}} U(\mathbf{M}_0, \widehat{\mathbf{M}}) = \nabla_{\widetilde{\mathbf{M}}} \boldsymbol{\Theta}'_\gamma (\mathbf{M}_0; \widetilde{\mathbf{M}}) \nabla_{\boldsymbol{\Theta}'} U(\mathbf{M}_0, \widehat{\mathbf{M}}). \tag{6.39}$$

Similarly to Eq. (6.35), the second gradient term $\nabla_{\boldsymbol{\Theta}'} U(\mathbf{M}_0, \widehat{\mathbf{M}})$ is relatively easier to evaluate. In the remainder of this section we focus on the computation of the first gradient term, which involves partial derivatives of $\boldsymbol{\Theta}' = (\mathbf{U}, \widetilde{\mathbf{U}}, \mathbf{V}, \boldsymbol{\Sigma})$ with respect to malicious users $\widetilde{\mathbf{M}}$.

We begin with the KKT condition at the optimal solution $\boldsymbol{\Theta}'$ of Eq. (6.29). Unlike the alternating minimization formulation, the nuclear norm function $\| \cdot \|_*$ is not everywhere differentiable. As a result, the KKT condition relates the *subdifferential* of the nuclear norm function $\partial \| \cdot \|_*$ as

$$\mathcal{R}_{\Omega, \widetilde{\Omega}} \left([\mathbf{M}_0; \widetilde{\mathbf{M}}] - [\mathbf{X}; \widetilde{\mathbf{X}}] \right) \in \gamma \partial \| [\mathbf{X}; \widetilde{\mathbf{X}}] \|_*. \tag{6.40}$$

Here $[\mathbf{X}; \widetilde{\mathbf{X}}]$ is the concatenated $(n + n') \times m$ matrix of $\mathbf{X}$ and $\widetilde{\mathbf{X}}$. The subdifferential of the nuclear norm function $\partial \| \cdot \|_*$ is also known [Candès and Recht, 2007]:

$$\partial \| \mathbf{X} \|_* = \left\{ \mathbf{U}\mathbf{V}^{\top} + \mathbf{W} : \mathbf{U}^{\top}\mathbf{W} = \mathbf{W}\mathbf{V} = 0, \| \mathbf{W} \|_2 \leq 1 \right\},$$

where $\mathbf{X} = \mathbf{U}\boldsymbol{\Sigma}\mathbf{V}^{\top}$ is the singular value decomposition of $\mathbf{X}$. Suppose $\{\boldsymbol{u}_i\}, \{\tilde{\boldsymbol{u}}_i\}$ and $\{\boldsymbol{v}_j\}$ are rows of $\mathbf{U}, \widetilde{\mathbf{U}}, \mathbf{V}$ and $\mathbf{W} = \{w_{ij}\}$. We can then re-formulate the KKT condition Eq. (6.40) as follows:

$$\begin{aligned} \forall (i, j) \in \Omega, \quad & \mathbf{M}_{0ij} = \boldsymbol{u}_i^{\top} (\boldsymbol{\Sigma} + \gamma \mathbf{I}_\rho) \boldsymbol{v}_j + \gamma w_{ij}; \\ \forall (i, j) \in \widetilde{\Omega}, \quad & \widetilde{\mathbf{M}}_{ij} = \tilde{\boldsymbol{u}}_i^{\top} (\boldsymbol{\Sigma} + \gamma \mathbf{I}_\rho) \boldsymbol{v}_j + \gamma \tilde{w}_{ij}. \end{aligned} \tag{6.41}$$

This enables us to derive $\nabla_{\widetilde{\mathbf{M}}} \boldsymbol{\Theta} = \nabla_{\widetilde{\mathbf{M}}} (\boldsymbol{u}, \tilde{\boldsymbol{u}}, \boldsymbol{v}, \sigma)$ (see Li et al. [2016] for further details).

Algorithm 6.6 Optimizing $\widetilde{\mathbf{M}}$ via SGLD

1: **Input:** Original partially observed $n \times m$ data matrix $\mathbf{M}_0$, algorithm regularization parameter γ, attack budget parameters α, C, and Λ, attacker's utility function R, step size $\{\beta_t\}_{t=1}^{\infty}$, tuning parameter β, number of SGLD iterations T.

2: **Prior setup:** compute $\xi_j = \frac{1}{m} \sum_{i=1}^{m} \mathbf{M}_{0ij}$ and $\sigma_j^2 = \frac{1}{m} \sum_{i=1}^{m} (\mathbf{M}_{0ij} - \xi_j)^2$ for every $j \in [n]$.

3: **Initialization:** sample $\widetilde{\mathbf{M}}_{ij}^{(0)} \sim \mathcal{N}(\xi_j, \sigma_j^2)$ for $i \in [m']$ and $j \in [n]$.

4: **for** $t = 0$ to T **do**

5:     Compute the optimal solution $\mathbf{\Theta}_{\gamma}(\mathbf{M}_0; \widetilde{\mathbf{M}}^{(t)})$.

6:     Compute gradient $\nabla_{\widetilde{\mathbf{M}}} U(\mathbf{M}_0, \widehat{\mathbf{M}})$ using Eq. (6.34).

7:     Update $\widetilde{\mathbf{M}}^{(t+1)}$ according to Eq. (6.45).

8: **end for**

9: **Projection:** find $\widetilde{\mathbf{M}}^* \in \arg\min_{\widetilde{\mathbf{M}} \in \mathcal{M}} \|\widetilde{\mathbf{M}} - \widetilde{\mathbf{M}}^{(t)}\|_F^2$.

10: **Output:** $n' \times m$ malicious matrix $\mathbf{M}^*$.

6.4.4   MIMICKING NORMAL USER BEHAVIORS

Normal users generally do not rate items uniformly at random. For example, some movies are significantly more popular than others. As a result, malicious users that pick rated movies uniformly at random can be easily identified by running a t-test against a known database consisting of only normal users. To alleviate this issue, this section describes an alternative approach to compute data poisoning attacks such that the resulting malicious users $\widetilde{\mathbf{M}}$ mimic normal users $\mathbf{M}_0$ to avoid potential detection, while still achieving reasonably high utility $U(\mathbf{M}_0, \widehat{\mathbf{M}})$ for the attacker. We use a Bayesian formulation to take both data poisoning and stealth objectives into consideration. The prior distribution $p_0(\widetilde{\mathbf{M}})$ captures normal user behaviors and is defined as a multivariate normal distribution

$$p_0(\widetilde{\mathbf{M}}) = \prod_{i=1}^{m'} \prod_{j=1}^{n} \mathcal{N}(\widetilde{\mathbf{M}}_{ij}; \xi_j, \sigma_j^2), \tag{6.42}$$

where ξ_j and σ_j^2 are mean and variance parameters for the rating of the jth item provided by normal users. In practice, both parameters can be estimated using normal user matrix $\mathbf{M}_0$ as $\xi_j = \frac{1}{m} \sum_{i=1}^{m} \mathbf{M}_{0ij}$ and $\sigma^2 = \frac{1}{n} \sum_{i=1}^{n} (\mathbf{M}_{0ij} - \xi_j)^2$. On the other hand, the likelihood $p(\mathbf{M}_0|\widetilde{\mathbf{M}})$ is defined as

$$p(\mathbf{M}_0|\widetilde{\mathbf{M}}) = \frac{1}{Z} \exp\left(\mu U(\mathbf{M}_0 \widehat{\mathbf{M}})\right), \tag{6.43}$$

where $U(\mathbf{M}_0, \widehat{\mathbf{M}}) = U(\widehat{\mathbf{M}}(\mathbf{\Theta}_{\gamma}(\mathbf{M}_0; \widetilde{\mathbf{M}})), \mathbf{M}_0)$ is one of the attacker utility models defined above (for example, corresponding to the reliability attack), Z is a normalization constant, and $\mu > 0$ is a tuning parameter that trades off attack performance and stealth. A small μ shifts the posterior of $\widetilde{\mathbf{M}}$ toward its prior, which makes the resulting attack strategy less effective but harder to detect, and vice versa.

Given both prior and likelihood functions, an effective stealthy attack strategy $\widetilde{\mathbf{M}}$ can be obtained by sampling from its posterior distribution:

$$p(\widetilde{\mathbf{M}}|\mathbf{M}_0) = p_0(\widetilde{\mathbf{M}})p(\mathbf{M}_0|\widetilde{\mathbf{M}})/p(\mathbf{M}_0)$$

$$\propto \exp\left(-\sum_{i=1}^{n'}\sum_{j=1}^{m}\frac{(\widetilde{\mathbf{M}}_{ij} - \xi_j)^2}{2\sigma_j^2} + \mu U(\mathbf{M}_0, \widehat{\mathbf{M}})\right). \tag{6.44}$$

Posterior sampling of Eq. (6.44) is intractable due to the implicit and complicated dependency of the estimated matrix $\widehat{\mathbf{M}}$ on the malicious data $\widetilde{\mathbf{M}}$, that is, $\widehat{\mathbf{M}} = \widehat{\mathbf{M}}(\Theta_\gamma(\mathbf{M}_0; \widetilde{\mathbf{M}}))$. To circumvent this problem, we can apply *Stochastic Gradient Langevin Dynamics (SGLD)* [Welling and Teh, 2011] to approximately sample $\widetilde{\mathbf{M}}$ from its posterior distribution in Eq. (6.44). More specfically, the SGLD algorithm iteratively computes a sequence of posterior samples $\{\widetilde{\mathbf{M}}^{(t)}\}_{t\geq 0}$ and in iteration t the new sample $\widetilde{\mathbf{M}}^{(t+1)}$ is computed as

$$\widetilde{\mathbf{M}}^{(t+1)} = \widetilde{\mathbf{M}}^{(t)} + \frac{\beta_t}{2}\left(\nabla_{\widetilde{\mathbf{M}}}\log p(\widetilde{\mathbf{M}}|\mathbf{M}_0)\right) + \varepsilon_t, \tag{6.45}$$

where $\{\beta_t\}_{t\geq 0}$ are step sizes and $\varepsilon_t \sim \mathcal{N}(\mathbf{0}, \beta_t\mathbf{I})$ are independent Gaussian noises injected at each SGLD iteration. The gradient $\nabla_{\widetilde{\mathbf{M}}}\log p(\widetilde{\mathbf{M}}|\mathbf{M}_0)$ can be computed as

$$\nabla_{\widetilde{\mathbf{M}}}\log p(\widetilde{\mathbf{M}}|\mathbf{M}_0) = -(\widetilde{\mathbf{M}} - \Xi)\Sigma^{-1} + \mu\nabla_{\widetilde{\mathbf{M}}}U(\mathbf{M}_0, \widehat{\mathbf{M}}),$$

where $\Sigma = \mathbf{diag}(\sigma_1^2, \cdots, \sigma_n^2)$ and Ξ is an $m' \times n$ matrix with $\Xi_{ij} = \xi_j$ for $i \in [m']$ and $j \in [m]$. The other gradient $\nabla_{\widetilde{\mathbf{M}}}U(\mathbf{M}_0, \widehat{\mathbf{M}})$ can be computed using the procedure in Sections 6.4.2 and 6.4.3. Finally, the sampled malicious matrix $\widetilde{\mathbf{M}}^{(t)}$ is projected back onto the feasible set $\mathbb{M}$ by selecting C items per user with the largest absolute rating and truncating ratings to the level of $\{\pm\Lambda\}$. A high-level description of this method is given in Algorithm 6.6.

6.5   A GENERAL FRAMEWORK FOR POISONING ATTACKS

We now describe a rather general approach for poisoning attacks that was introduced by Mei and Zhu [2015a], and connected to the problem of *machine teaching*. This approach allows for both the possibility of adding and modifying a collection of data which subsequently becomes a part of the training data set. Moreover, it can, in principle, be applied in both supervised and unsupervised learning settings, although for our description of this approach below it's most natural to consider a supervised learning problem.

Let's start with the traditional learning problem which computes an optimal parametrization w and trades off empirical risk and a regularization term. Let $\mathcal{D}$ be a dataset. Since the poisoning attack modifies this dataset, we now explicitly represent everything (including w) as a function of training data (original or modified). The traditional learning problem can then be formulated as

$$w(\mathcal{D}) \in \arg\max_{w} \sum_{i \in \mathcal{D}} l_i(w) + \gamma \rho(w), \tag{6.46}$$

where $l_i(w)$ is the loss on a datapoint i and $\rho(w)$ the regularization term. For example, in binary classification the loss could be $l_i(w) = l(y_i g(x_i; w))$ for the label y_i and classification score $g(x_i; w)$. Suppose that both $\rho(w)$ and $l_i(w)$ are strictly convex and twice continuously differentiable. Mei and Zhu [2015a] consider the more general case by allowing constraints as a part of the learner's optimization problem, but our restriction allows a simpler presentation of the approach.

Again, to distinguish the original "clean" dataset from the one produced by an attacker, we let $\mathcal{D}_0$ denote the former, while $\mathcal{D}$ represents the latter. The attacker's decision is then to create a new dataset $\mathcal{D}$, starting with $\mathcal{D}_0$. In doing so, the attacker faces the tradeoff we described in the beginning of the chapter: on the one hand the attacker wishes to minimize its own risk function, $R_A(w(\mathcal{D}))$, where $w(\mathcal{D})$ is the parameter produced by the learner when the training dataset $\mathcal{D}$ is used; on the other hand, the attacker incurs a cost captured by the cost function $c(\mathcal{D}_0, \mathcal{D})$. Mei and Zhu consider the following optimization problem as a formalization of this tradeoff:

$$\min_{\mathcal{D}} \quad R_A(w(\mathcal{D})) + c(\mathcal{D}_0, \mathcal{D})$$
$$\text{s.t.:} \quad w(\mathcal{D}) \in \arg\max_{w} \sum_{i \in \mathcal{D}} l_i(w) + \gamma \rho(w). \tag{6.47}$$

This is, of course, a challenging bi-level optimization problem. However, note that if the learner's problem is strictly convex, we can rewrite the constraint using the corresponding first-order conditions:

$$\forall j, \sum_{i \in \mathcal{D}} \frac{\partial l_i(w)}{\partial w_j} + \gamma \frac{\partial \rho(w)}{\partial w_j} = 0. \tag{6.48}$$

Suppose that $\mathbb{D}$ is the space of all feasible datasets that the attacker can generate (for example, the attacker cannot delete data from $\mathcal{D}_0$). We can then, in principle, optimize the attacker's objective using projected gradient descent, where the update in iteration $t + 1$ is

$$\mathcal{D}^{t+1} = \text{Proj}_{\mathbb{D}} \left[\mathcal{D}^t - \beta_t \nabla_{\mathcal{D}} R_A(w(\mathcal{D})) - \nabla_{\mathcal{D}} c(\mathcal{D}, \mathcal{D}_0) \right], \tag{6.49}$$

and where β_t is the learning rate. $\nabla_{\mathcal{D}} c(\mathcal{D}, \mathcal{D}_0)$ can be computed directly from the analytic expression of $c(\cdot)$, and

$$\nabla_{\mathcal{D}} R_A(w(\mathcal{D})) = \nabla_w R_A(w) \frac{\partial w}{\partial \mathcal{D}}. \tag{6.50}$$

While $\nabla_w R_A(w)$ is also available directly from the analytic form of the attacker's risk function, $\frac{\partial w}{\partial \mathcal{D}}$ is implicitly represented by the first-order conditions (which we presented as constraints above).

Fortunately, we can take advantage of the implicit function theorem, which (under the conditions we mention presently) allows us to compute this derivative. First, define the collection of functions

$$f_j(\mathcal{D}, w) = \sum_{i \in \mathcal{D}} \frac{\partial l_i(w)}{\partial w_j} + \gamma \frac{\partial \rho(w)}{\partial w_j}. \tag{6.51}$$

We can collect these into a vector $f(\mathcal{D}, w)$, noting that $f(\mathcal{D}, w) = 0$ represents the first-order conditions above. Let

$$\frac{\partial f_j}{\partial w_k} = \sum_{i \in \mathcal{D}} \frac{\partial^2 l_i(w)}{\partial w_j \partial w_k} + \gamma \frac{\partial^2 \rho(w)}{\partial w_j \partial w_k}, \tag{6.52}$$

i.e., the Hessian of the original optimization problem for the learner, which we denote in the corresponding matrix form simply by $\frac{\partial f}{\partial w}$, and let $\frac{\partial f_j}{\partial \mathcal{D}}$ be the partial derivative of f with respect to the dataset $\mathcal{D}$ (we illustrate how this can be computed more concretely below); we denote the corresponding matrix by $\partial f / \partial \mathcal{D}$. Then, if $\frac{\partial f}{\partial w}$ is full rank (and therefore has an inverse), we can compute

$$\frac{\partial w}{\partial \mathcal{D}} = - \left[\left[\frac{\partial f}{\partial w} \right]^{-1} \frac{\partial f}{\partial \mathcal{D}} \right]. \tag{6.53}$$

To make things concrete, suppose that the attacker is attacking a logistic regression, and is only able to modify the feature vectors in the dataset $\mathcal{D}_0$. Thus, the attacker's decision is to compute a new feature matrix $\mathbf{X}$, given the original feature matrix $\mathbf{X}_0$ and original binary labels y_0. Let $c(\mathbf{X}_0, \mathbf{X}) = \|\mathbf{X} - \mathbf{X}_0\|_F$, the Frobenius norm of the difference between the attack and original feature matrices. Consider a *targeted* attack with a target parameter vector w_T that the attacker wishes the defender to learn. Thus, we let $R_A(w) = \|w - w_T\|_2^2$.

Since the attacker is modifying only the existing feature vectors, the partial derivative $\partial f / \partial \mathcal{D}$ is composed of partial derivatives with respect to corresponding features k of each datapoint i, $\partial f_j / \partial x_{ik}$. Since the regularizer does not depend on x, we omit this term (as it becomes zero), and focus on the loss term. The logistic loss function is $l_i(w) = l(y_i g(x_i; w)) = -\log(\sigma(y_i g_i))$, where $g_i = w^T x_i + b$ and $\sigma(a) = 1/(1 + e^{-a})$ is the logistic function. A useful fact is that the first derivative of the logistic function is $\sigma'(a) = \sigma(a)(1 - \sigma(a))$. Then, observe that in this case,

$$f_j(\mathbf{X}, w) = -\sum_i (1 - \sigma(y_i g_i)) y_i x_{ij}. \tag{6.54}$$

Consequently,

$$\frac{\partial f_j}{\partial x_{ik}} = \sigma(y_i g_i)(1 - \sigma(y_i g_i)) y_i x_{ij} w_k - (1 - \sigma(y_i g_i)) y_i \mathbb{1}(j = k), \tag{6.55}$$

where $\mathbb{1}(j = k)$ is the identity function which is 1 if $j = k$ and 0 otherwise.

6.6   BLACK-BOX POISONING ATTACKS

All of our discussion of poisoning attacks in the preceding sections assumed that the attacker knows everything there is to know about the system they are attacking—in other words, these were white-box attacks. We now turn to the question of how feasible it is to deploy poisoning attacks without such detailed knowledge. We therefore consider again the problem of *black-box attacks*, now in the context of data poisoning. To be more precise, there are three pieces of information that the attacker needs to know in a white-box poisoning attack: a feature space, F, the dataset that the learner would have used before the poisoning attack, D (note that we indicate by D the knowledge abou the dataset, rather than the dataset itself), and the algorithm A used by the learner (including relevant hyperparameters).

Figure 6.2 presents an information lattice for black-box poisoning attacks, starting with a white-box attack (full information). If the learner does not know the algorithm, a black-box attack can use a proxy algorithm instead, and evaluate robustness of poisoning attacks with respect to incorrect assumptions about the algorithm being used. If feature space is unknown, a proxy feature space may also be used, although this is a severe limitation on the information that the attacker possesses. However, the most significant limitation may be poor information about the dataset being poisoned. If the proxy data is only partial data used by the learner, it is still possible for the attacker to modify the instances (including labels) in the dataset (in principle, anyway), but such attacks are clearly impossible without having *some* access to the training data. Insertion attacks, on the other hand, can still be introduced, although their effectiveness is sure to degrade if proxy data is not representative of actual training data used by the learner.

6.7   BIBLIOGRAPHIC NOTES

The problem of learning with noise has a long tradition in machine learning [Bshoutya et al., 2002, Kearns and Li, 1993, Natarajan et al., 2013]. However, these are focused on worst-case errors for a small number of samples, rather than specific algorithms for adversarial data poisoning. We tackle the issue of robust learning in the presence of adversarial poisoning attacks in the next chapter.

Some of the earliest formal models and algorithmic approaches to poisoning attacks were label-flipping attacks against binary classifiers (primarily, linear SVM). Our description is based on Xiao et al. [2012]. An alternative approach is presented by Biggio et al. [2011], and Xiao et al. [2015] present a unified treatment of both.

The poisoning attack on support vector machines in which a single malicious feature vector is added is due to Biggio et al. [2012]. This attack is rather restricted: only a single instance is added to the data, and the adversary has no control over the label (for example, the adversary may perform a carefully designed malicious or benign task, such as sending an email to the recipient in an organization, but the learner subsequently ensures that these are labeled correctly for training). If an adversary is able to add more than a single malicious instance to the training

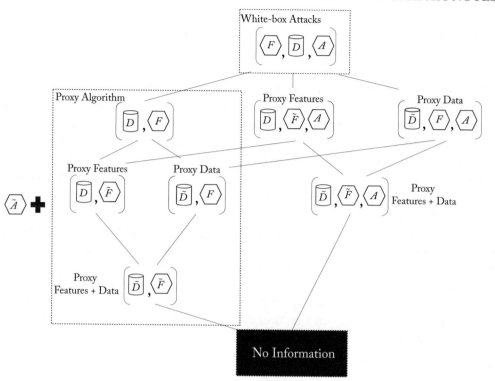

Figure 6.2: An lattice of black-box data poisoning attacks on machine learning.

data, they could apply the approach we described in a greedy fashion, adding one datapoint at a time. A more general approach based on machine teaching (see below), on the other hand, considers the impact of modifying the entire dataset.

In the unsupervised problem space, several approaches consider poisoning attacks on clustering methods [Biggio et al., 2014a,b], and several deal with poisoning anomaly detectors [Kloft and Laskov, 2012, Rubinstein et al., 2009]. As discussed above, the attack on conventional centroid anomaly detection methods (which are quite generally used) is mathematically relatively straightforward, and is due to Kloft and Laskov [2012]; however, we know of no existing attacks on kernel-based centroid anomaly detectors. The attack on PCA-based anomaly detection is due to Rubinstein et al. [2009]. Finally, the attacks on matrix completion (for example, as used in recommender systems) are due to Li et al. [2016].

The general approach for poisoning attacks, as well as the connection to machine teaching, are due to Mei and Zhu [2015a]. A similar approach within the machine teaching framework has also been explored by Mei and Zhu [2015b] in a specific attack on Latent Dirichlet Allocation (LDA), particularly in the context of natural language topic modeling. Another recent general approach to data poisoning in supervised learning settings is due to Koh and Liang [2017], who use influence functions to study the impact of small perturbations in training datapoints on learning, and apply this idea to poisoning deep learning. While a large portion of methods for poisoning target classifiers, a recent attack considers linear regression [Jagielski et al., 2018].

Our discussion of black-box approaches for poisoning attacks is closely connected to a recent framework for categorizing information possessed by the attacker due to Suciu et al. [2018]. They term this the *FAIL* framework, which we discussed earlier (Chapter 3, bibliographic notes), where F corresponds to knowledge about feature space, A refers to knowledge about the algorithm, I is what we call knowledge about data (they term it *instances*), and L is a reference to the attacker's capability, which we here treat orthogonally (in their example, this refers to which features the attacker can modify, an issue we mostly do not address, but this can also consider limitations on the attacker such as what the data they can modify). Suciu et al. [2018] also develop an effective targeted black-box poisoning attack algorithm, *StingRay*. The high-level idea behind *StingRay* is to use a collection of base datapoints (e.g., from the actual or proxy dataset) which are labeled with a target label and are close to the target feature vector in feature space. The adversary then modifies features in the base instance to move the associated feature vector closer to the target. *StingRay* introduces several additional considerations into its attack: first, it also attempts to minimize impact on other instances which are not the target (the stealth consideration), and second, it ensures that it is not pruned by a detector which is meant to sanitize the training data.

CHAPTER 7

Defending Against Data Poisoning

Making machine learning algorithms robust against malicious noise in training data is one of the classic problems in machine learning. We define this *robust learning problem* as follows. We start with the pristine training dataset $\mathcal{D}_0$ of n labeled examples. Suppose that an unknown proportion α of the dataset $\mathcal{D}_0$ is then corrupted arbitrarily (i.e., both feature vectors and labels may be corrupted), resulting in a corrupted dataset $\mathcal{D}$. The goal is to learn a model f on the corrupted data $\mathcal{D}$ which is nearly as good (in terms of, say, prediction accuracy) as a model f_0 learned on pristine data $\mathcal{D}_0$.

We divide the algorithmic approaches for poisoning-robust learning into three categories.

1. **Data sub-sampling:** take many random sub-samples of $\mathcal{D}$, learn a model on each using the same learning algorithm, and choose the model with the smallest (training) error (e.g., Kearns and Li [1993]).

2. **Outlier removal:** identify and remove anomalous instances (outliers), and then learn the model (e.g., Klivans et al. [2009]).

3. **Trimmed optimization:** (arguably, a variation on theme 2) minimize empirical risk while pruning out the $(1 - \alpha)n$ datapoints with the largest error (e.g., Liu et al. [2017]).

In this chapter, we present example methods for defending against poisoning attacks for each of these categories.

7.1 ROBUST LEARNING THROUGH DATA SUB-SAMPLING

The first approach we present is also one of the oldest. In a seminal paper, Kearns and Li [1993] present one of the earliest approaches for robust classification. The key idea is that, if α is sufficiently small compared to target error ϵ, *any* polynomial time PAC learning algorithm can be used (as a subroutine) to obtain a PAC learning algorithm when a fraction α of data has malicious noise.

The algorithm is given in Algorithm 7.7. This algorithm is actually slightly different from the original idea by Kearns and Li [1993]: originally, one would take K samples of size m from

---

Algorithm 7.7 Data Sub-sampling Algorithm

---

    for $i = 1 \cdots K$ **do**
      $\mathcal{D}_i = \text{Sample}(\mathcal{D}, m)$
      $h_i = \text{Learn}(\mathcal{D}_i)$
      $e_i = \text{Error}(h_i, \mathcal{D}_i)$
    end for
    $i^* = \arg\min_i e_i$
    return h_{i^*}.

---

an oracle, which generates malicious instances with probability α. In contrast, we use a setup that has become more conventional, starting with a dataset $\mathcal{D}$, which has (at most) a fraction α of poisoned instances. Then, we take K subsamples of size m from this dataset (which we assume is sufficiently large, say, with at least Km instances). The Sample() function takes one such subsample. Next, the Learn() function applies a learning algorithm. Finally, we measure the *training* error e_i of the hypothesis h_i returned by the (non-robust) learning algorithm.

After K steps, the algorithm simply returns the hypothesis which obtained the smallest training error. The key insight is that when the fraction of data poisoned, α, is very small, it is very likely that one of the K samples contains no malicious samples, which would allow a *good* learning algorithm to obtain a small training error. The following theorem formalizes this (see Chapter 2 for the formal definition of a *polynomial-time PAC learning algorithm* in this theorem).

Theorem 7.1   *Suppose Learn() implements a polynomial-time PAC learning algorithm, and m is its sample complexity to achieve error $\epsilon/2$ with probability at least $1/2$. Let $\alpha \leq (\log m)/m$, and $K \geq 2m^2 \log(3/\delta)$. Then, with probability at least $1 - \delta$ the (true) error of the solution computed by Algorithm 7.7 is at most ϵ.*

While Algorithm 7.7 and the theoretical guarantee based on it are specific to binary classification, the algorithm itself is not difficult to generalize: in fact, we can simply replace the function which computes error with any measure of risk on data, and the approach would directly extend to regression or, for that matter, to unsupervised learning which aims to minimize some measure of empirical risk (such as maximizing likelihood of data).

7.2   ROBUST LEARNING THROUGH OUTLIER REMOVAL

The next general technique for dealing with poisoned data is to attempt to identify and remove the malicious instances from the training data before learning. At the high level, approaches of this type work as described in Algorithm 7.8.

Klivans et al. [2009] introduce a formal learning framework based on outlier detection and removal. Suppose that the model class $\mathcal{F}$ is the class of origin-centered linear classifiers, i.e.,

---

Algorithm 7.8 Robust Learning Using Outlier Removal

Input: dataset $\mathcal{D}$

$\mathcal{D}_{clean} = \text{RemoveOutliers}(\mathcal{D})$

$h = \text{Learn}(\mathcal{D}_{clean})$

return h

$f(x) = \text{sgn}\{w^T x\}$. Klivans et al. [2009] then propose an algorithm for outlier removal based on PCA, which leads to PAC-learning guarantees for a particular instance of Learn() described below, albeit under relatively strong distributional assumptions.

Specifically, consider the RemoveOutliers() function. Klivans et al. [2009] suggest the iterative approach for removing outliers in Algorithm 7.9. Intuitively, this algorithm iteratively

---

Algorithm 7.9 RemoveOutliers()

Input: dataset $\mathcal{D}$

$\mathcal{D}_{clean} = \mathcal{D}$

repeat

    Define $A = \sum_{x \in \mathcal{D}_{clean}} xx^T$.

    Find v, the eigenvector with the largest eigenvalue of A.

    S: the set of feature vectors $x \in \mathcal{D}$ with $(v^T x)^2 \geq \frac{10 \log n}{m}$.

    $\mathcal{D}_{clean} \leftarrow \mathcal{D} - S$

until $S = \emptyset$

return $\mathcal{D}$.

---

projects the data into a single dimension with the highest variance, and removes all outliers along this dimension. Their Learn() algorithm is just a simple averaging approach, which computes a weight vector of a linear classifier as

$$w = \frac{1}{|\mathcal{D}_{clean}|} \sum_{i \in \mathcal{D}_{clean}} y_i x_i. \tag{7.1}$$

This clearly results in a polynomial time algorithm. With this, they were able to prove the following result.

Theorem 7.2 *Suppose that the distribution over feature vectors is uniform over a unit ball, and malicious noise $\alpha \leq \Omega(\epsilon^2 / \log(n/\epsilon))$. Then, the algorithm above learns an origin-centered linear classifier with accuracy of at least $1 - \epsilon$.*

Again, while the algorithm due to Klivans et al. [2009], and the theoretical guarantee, are specific to linear classification, the idea is not difficult to generalize. Indeed, any other outlier detection approach may in principle be used, and in any case, once a clean dataset is obtained, we can apply an arbitrary learning algorithm to it (of course, the theoretical result is specific to linear classification).

Algorithm 7.10 Robust Learning with Micromodels

Input: dataset $\mathcal{D}$, number of micromodels K

$\mathcal{D}_{clean} \leftarrow \emptyset$

$\{\mathcal{D}_i\} = \text{PartitionData}(\mathcal{D}, K)$

for $i = 1$ to K **do**

　$h_i = AD(\mathcal{D}_i)$

end for

for $x \in D$ **do**

　$s(x) = \sum_i w_i h_i(x)$

　if $s(x) \leq r$ **then**

　　$\mathcal{D}_{clean} \leftarrow \mathcal{D}_{clean} \cup x$

　end if

end for

$h = \text{Learn}(\mathcal{D}_{clean})$

return h

Another approach for outlier removal, suggested by Cretu et al. [2008], makes use of anomaly detection, and is actually designed specifically for robust anomaly detection in security. Let $AD(\mathcal{D})$ be an anomaly detector which takes a dataset $\mathcal{D}$ as input and returns a model $f(x)$ which outputs *normal* (-1) and *anomalous* (+1) for an arbitrary input x. Now, suppose we partition the dataset $\mathcal{D}$ into a collection of subsets $\{\mathcal{D}_i\}$, and train an anomaly detector independently for each $\mathcal{D}_i$. This provides us with a collection of detectors, $\{h_i\}$, which Cretu et al. [2008] term *micromodels*. We can now use the ensemble of h_i to score each datapoint in $\mathcal{D}$ as normal or anomalous. Specifically, for each $x \in \mathcal{D}$, let the score $s(x) = \sum_i w_i h_i(x)$ be the weighted vote on this datapoint by all micromodels. We then remove x from $\mathcal{D}$ if $s(x) \geq r$, for some predefined threshold r. Once the data has thereby been sanitized, we can learn the final model (which could be a classifier, regression, or an anomaly detector) on the sanitized dataset.[1] The full algorithm for cleaning the data using micromodels is given by Algorithm 7.10.

A third idea for outlier removal as a means to sanitize data is due to Barreno et al. [2010], and can be viewed as a variation on the notion of micromodels. Barreno et al. [2010] assume that they start with a pristine training dataset $\mathcal{D}_\star$, and consider adding an additional dataset $\mathcal{Z}$ which may be partially poisoned. At the high level, they evaluate how much impact each

---

[1]We note that Cretu et al. [2008] actually suggest splitting the original dataset into three parts: the first to learn the micromodels, the second which is sanitized and then used to learn the anomaly detector (or, in our case, any other learning model), and the third for evaluation. We reframe their methodology so it can be used to sanitize the training data directly (which is also used to learn the micromodels).

datapoint $z \in \mathcal{Z}$ has on the marginal change in empirical risk of a learned model, an approach that Barreno et al. [2010] term *Reject on Negative Impact* or *RONI*.

Algorithm 7.11 RONI Algorithm

Input: pristine dataset $\mathcal{D}_\star$, new dataset $\mathcal{Z}$
$\mathcal{D} \leftarrow \mathcal{D}_\star$
$\mathcal{C} = \text{Sample}(\mathcal{D}_\star)$
$\{(\mathcal{T}_i, \mathcal{Q}_i)\} = \text{PartitionData}(\mathcal{D}_\star - \mathcal{C}, K)$
for $c \in \mathcal{C}$ **do**
  $s(c) = \text{FindShift}(c, \{(\mathcal{T}_i, \mathcal{Q}_i)\})$
end for
$a = \text{Average}(\{s(c)\})$
for $z \in \mathcal{Z}$ **do**
  $s(z) = \text{FindShift}(z, \{(\mathcal{T}_i, \mathcal{Q}_i)\})$
  if $s(z) \geq 0$ or $a - s(z) \leq r$ **then**
    $\mathcal{D} \leftarrow \mathcal{D} \cup z$
  end if
end for
return $\mathcal{D}$

Algorithm 7.12 FindShift()

---

for $i = 1$ to K **do**
  $h_i = \text{Learn}(\mathcal{T}_i)$
  $\tilde{h}_i = \text{Learn}(\mathcal{T}_i \cup z)$
  $e_i = \text{Error}(h_i, \mathcal{Q}_i)$
  $\tilde{e}_i = \text{Error}(\tilde{h}_i, \mathcal{Q}_i)$
end for
$e_{ave} = \text{Average}(\{e_i\})$
$\tilde{e}_{ave} = \text{Average}(\{\tilde{e}_i\})$
return $e_{ave} - \tilde{e}_{ave}$

Algorithm 7.11 presents the full RONI approach, where the function FindShift(), elaborated in Algorithm 7.12, returns the impact of a given datapoint z on average accuracy (equivalently, error) over the collection of training and test subsamples.

The first step in RONI is to randomly sample a calibration dataset $\mathcal{C}$ from $\mathcal{D}_\star$. Next, we split the dataset $\mathcal{D}_\star - \mathcal{C}$ into K randomly sampled pairs of training and test subsets, $\mathcal{T}_i$ and $\mathcal{Q}_i$, respectively. We then take the average impact on accuracy that each $c \in \mathcal{C}$ has as a baseline (since we assume that $\mathcal{C}$ is sampled from pristine data). Next, we iterative score each $z \in \mathcal{Z}$

in a similar fashion in terms of its average impact on accuracy over the collection of $\mathcal{T}_i$ (used for training) and $\mathcal{Q}_i$ (used for evaluation) pairs. Finally, we filter any z which has a sufficiently large (anomalous) negative impact on learning accuracy as compared to the baseline, where an exogenously specified threshold r determines how conservative we are at doing such filtering.

7.3   ROBUST LEARNING THROUGH TRIMMED OPTIMIZATION

We now illustrate the third approach for making machine learning robust to training data poisoning, *trimmed optimization*, in the context of linear regression learning.

The specific approach we describe actually combines linear regression with PCA, and the full approach is given in Algorithm 7.13. In this algorithm, the first step performs PCA, and

---
Algorithm 7.13 Robust Principle Component Regression

Input: dataset $\mathcal{D}$
$\mathbf{B} = \text{findBasis}(\mathcal{D})$
$w = \text{learnLinearRegression}(\mathcal{D}, \mathbf{B})$,

---

the second learns the actual linear regression, using the PCA basis $\mathbf{B}$. It is clear that both steps need to be performed robustly. Right now, we focus on step 2, and assume that the basis $\mathbf{B}$ is computed correctly. We deal with robust PCA in Section 7.4.

We now formalize the setup. We start with the pristine training dataset $\mathcal{D}_\star$ of n labeled examples, $\langle \mathbf{X}_\star, y_\star \rangle$, where $y_\star \in \mathbb{R}$, which subsequently suffers from two types of corruption: noise is added to feature vectors, and the adversary adds n_1 malicious examples (feature vectors and labels) to mislead learning. Thus, $\alpha = n_1/(n + n_1)$, and we define $\gamma = \frac{n_1}{n}$ as the *corruption ratio*, or the ratio of corrupted and pristine data. We assume that the adversary has full knowledge of the learning algorithm. The learner's goal is to learn a model on the corrupted dataset which is similar to the true model. We assume that $\mathbf{X}_\star$ is low-rank with a basis $\mathbf{B}$, and we assume that the true model is the associated low-dimensional linear regression.

Formally, observed training data is generated as follows.

1. **Ground truth:** $y_\star = \mathbf{X}_\star w^\star = \mathbf{U} w_U^\star$, where $w^\star$ is the true model weight vector, $w_U^\star$ is its low-dimensional representation, and $\mathbf{U} = \mathbf{X}_\star \mathbf{B}$ is the low-dimensional embedding of $\mathbf{X}_\star$.

2. **Noise:** $\mathbf{X}_0 = \mathbf{X}_\star + \mathbf{N}$, where $\mathbf{N}$ is a noise matrix with $\|\mathbf{N}\|_\infty \leq \epsilon$; $y_0 = y_\star + e$, where e is i.i.d. zero-mean Gaussian noise with variance σ.

3. **Corruption:** The attacker adds n_1 adversarially crafted datapoints $\{x_a, y_a\}$ to get $\langle \mathbf{X}, y \rangle$, which maximally skews prediction performance of low-dimensional linear regression.

We now present the trimmed regression algorithm proposed by Liu et al. [2017]. To estimate $\mathbf{y} = \mathbf{X}_\star w + e$, we assume $w_U = \mathbf{B}w$. Since $\mathbf{X}_\star = \mathbf{U}_\star \mathbf{B}$, we convert the estimation problem of w from a high dimensional space to the estimation problem of w_U in a low dimensional space, such that $y = \mathbf{U}w_U + e$. After computing an estimate $\widehat{w_U}$, we can convert it back to get $\widehat{w} = \mathbf{B}\widehat{w_U}$. Notice that this is analogous to standard principal component regression Jolliffe [1982]. However, the adversary may corrupt n_1 rows in $\mathbf{U}$ to fool the learner to cause wrong estimation on $\widehat{w_U}$, and thus on $\widehat{w}$. The trimmed regression algorithm (Algorithm 7.14) addresses this problem.

Algorithm 7.14 Trimmed Principal Component Regression

Input: X, B, y

1. Project $\mathbf{X}$ onto the span space of $\mathbf{B}$ and get $\mathbf{U} \leftarrow \mathbf{XB}^T$.

2. Solve the following minimization problem to get $\widehat{w_U}$

$$\min_{w_U} \sum_{j=1}^{n} \{(y_i - u_i w_U)^2 \text{ for } i = 1, ..., n + n_1\}_{(j)} \tag{7.2}$$

   where $z_{(j)}$ denotes the j-th smallest element in sequence z.

3. **return** $\widehat{w} \leftarrow \mathbf{B}\widehat{w_U}$.

Intuitively, during the training process we trim out the top n_1 samples that maximize the difference between the observed response y_i and the predicted response $u_i w_U$, where u_i denotes the i-th row of U. Since we know the variances of these differences are small (i.e., recall that σ is the variance of the random noise $y - xw^\star$), these samples corresponding to the largest differences are more likely to be the adversarial ones. Trimmed optimization problems of this kind appear at first to be quite intractable. In Section 7.5 we describe a scalable approach for solving such problems.

Liu et al. [2017] prove the following result.

Theorem 7.3 *Suppose that the basis $\mathbf{B}$ is given. Algorithm 7.14 returns $\widehat{w}$, such that for any real value $h > 1$ and for some constant c we have*

$$E_x[(x(\widehat{w} - w^\star))^2] \leq 4\sigma^2 \left(1 + \sqrt{\frac{1}{1-\gamma}}\right)^2 \log c \tag{7.3}$$

with probability at least $1 - c \cdot h^{-2}$.

Note that we can use a similar approach for robust learning more generally. The key idea above is to trim the risk function to exclude n_1 outliers. While in context this was specialized to the l_2 regression loss without regularization, one can immediately consider a more general version of the problem, where we minimize an arbitrary regularized risk, with a loss function $l(y, w^T x)$ and l_p regularization. Then, we can consider solving the following trimmed optimization problem to achieve robustness:

$$\min_{w} \sum_{j=1}^{n} \{l(y_i, w^T x_i) + \lambda \|w\|_p^p \text{ for } i = 1, ..., n + n_1\}_{(j)}, \tag{7.4}$$

where $z_{(j)}$, again, denotes the j-th smallest element in sequence z. As an example, we can apply this idea to robust classification with loss functions of the form $l(y w^T x)$.

7.4　ROBUST MATRIX FACTORIZATION

In this section, we discuss the approach by Liu et al. [2017] for recovering the low-rank subspace of matrix. While this is of independent importance, it would also allow us to solve the robust regression problem above.

Suppose that the observed and corrupted matrix $\mathbf{X}$ is generated as follows.

1. **Ground truth:** $\mathbf{X}_\star$ is the true low-rank matrix with a basis $\mathbf{B}$.

2. **Noise:** $\mathbf{X}_0 = \mathbf{X}_\star + \mathbf{N}$, where $\mathbf{N}$ is a noise matrix with $\|\mathbf{N}\|_\infty \leq \epsilon$.

3. **Corruption:** The attacker adds n_1 adversarially crafted rows $\{x_a\}$ to get the observed matrix $\mathbf{X}$.

The goal is to recover the true basis $\mathbf{B}$ of $\mathbf{X}_\star$. For convenience, we let $\mathcal{O}$ denote the set of (unknown) indices of the samples in $\mathbf{X}$ coming from $\mathbf{X}_0$ and $\mathcal{A} = \{1, ..., n + n_1\} - \mathcal{O}$ the set of indices for adversarial samples in $\mathbf{X}$. For an index set $\mathcal{I}$ and matrix $\mathbf{M}$, $\mathbf{M}^{\mathcal{I}}$ denotes the submatrix containing only rows in $\mathcal{I}$; similar notation is used for vectors.

7.4.1　NOISE-FREE SUBSPACE RECOVERY

We first consider an easier version of the robust subspace recorvery problem with $\mathbf{N} = 0$ (that is, no random noise is added to the matrix $\mathbf{X}_\star$; however, there are still n_1 malicious instances). In this case, we know that $\mathbf{X}^{\mathcal{O}} = \mathbf{X}_\star$. We assume that we know $\text{rank}(\mathbf{X}_\star) = k$ (or have an upper bound on it). Presently we show that there exists a sharp threshold θ on n_1 such that whenever $n_1 < \theta$, we can recover the basis $\mathbf{B}$ exactly with high probability, whereas if $n_1 \geq \theta$, the basis cannot be recovered. To characterize this threshold, we define the cardinality of the *maximal rank $k - 1$ subspace $MS_{k-1}(\mathbf{X}_\star)$* as the optimal value of the following problem:

$$\max_{\mathcal{I}} |\mathcal{I}| \text{ s.t. } \text{rank}(\mathbf{X}_\star^{\mathcal{I}}) \leq k - 1. \tag{7.5}$$

Intuitively, the adversary can insert $n_1 = n - MS_{k-1}(X_\star)$ samples to form a rank k subspace, which does not span $X_\star$. The following theorem shows that in this case, there is indeed no learner that can successfully recover the subspace of $X_\star$.

Theorem 7.4   *If $n_1 + MS_{k-1}(X_\star) \geq n$, then there exists an adversary such that no algorithm can recover the basis B with probability $> 1/2$.*

On the other hand, when n_1 is below this threshold, we can use Algorithm 7.15 to recover the subspace of $X_\star$.

Algorithm 7.15 Exact Recover Algorithm for Basis Recovery (Noisy-free)

We search for a subset $\mathcal{I}$ of indices, such that $|\mathcal{I}| = n$, and $\text{rank}(X^{\mathcal{I}}) = k$
return a basis of $X^{\mathcal{I}}$.

Theorem 7.5   *If $n_1 + MS_{k-1}(X_\star) < n$, then Algorithm 7.15 recovers B for any adversary.*

Theorems 7.4 and 7.5 together give the necessary and sufficient conditions for exact basis recovery. It can be show that $MS_{k-1}(X_\star) \geq k - 1$. Combining this with Theorem 7.4, we obtain the following upper bound on γ.

Corollary 7.6   *If $\gamma \geq 1 - \frac{k-1}{n}$, then we can successfully recover the basis B.*

7.4.2   DEALING WITH NOISE

We now consider the problem of robust PCA (basis recovery) when there is noise added to the true matrix $X_\star$. Clearly, in order for us to recover the basis when there is malicious noise we need to make sure that the problem can be solved even when no malicious noise is present. A sufficient condition, which is subsequently imposed, is that $X_\star$ is the unique optimal solution to the following problem:

$$\min_{X'} ||X_0 - X'||$$
$$\text{s.t. } \text{rank}(X') \leq k. \tag{7.6}$$

Note that this assumption is implied by the classical PCA problem [Eckart and Young, 1936, Hotelling, 1933, Jolliffe, 2002].

Unless otherwise mentioned, we use $|| \cdot ||$ to denote the Frobenius norm. We put no additional restrictions on additive noise N except above. We focus on the optimal value of the above problem, which we term the *noise residual* and denote by $NR(X_0) = N$. Noise residual is a key component to characterize the necessary and sufficient conditions for exact basis recovery with noise.

Characterization of the defender's ability to accurately recover the true basis $\mathbf{B}$ of $\mathbf{X}_\star$ after the attacker adds n_1 malicious instances stems from the attacker's ability to mislead the defender into thinking that some other basis, $\bar{\mathbf{B}}$, better represents $\mathbf{X}_\star$. Intuitively, since the defender does not know $\mathbf{X}_0$, $\mathbf{X}_\star$, or which n_1 rows of the data matrix $\mathbf{X}$ are adversarial, this comes down to the ability to identify the $n - n_1$ rows that correspond to the correct basis (note that it will suffice to obtain the correct basis even if some adversarial rows are used, since the adversary may be forced to align malicious examples with the correct basis to evade explicit detection). As we show below, whether the defender can succeed is determined by the relationship between the noise residual $NR(\mathbf{X}_0)$ and *sub-matrix residual*, denoted as $SR(\mathbf{X}_0)$, which is the value optimizing the following problem:

$$\min_{\mathcal{I},\mathbf{B},\mathbf{U}} ||\mathbf{X}_0^{\mathcal{I}} - \mathbf{U}\bar{\mathbf{B}}|| \tag{7.7a}$$

$$\text{s.t.} \quad \text{rank}(\bar{\mathbf{B}}) = k, \bar{\mathbf{B}}\bar{\mathbf{B}}^T = I_k, \mathbf{X}_\star\bar{\mathbf{B}}^T\bar{\mathbf{B}} \neq \mathbf{X}_\star \tag{7.7b}$$

$$\mathcal{I} \subseteq \{1, 2, ..., n\}, |\mathcal{I}| = n - n_1. \tag{7.7c}$$

We now explain the above optimization problem. $\mathbf{U}$ and $\bar{\mathbf{B}}$ are $(n - n_1) \times k$ and $k \times m$ matrices separately. Here $\bar{\mathbf{B}}$ is a basis which the attacker "targets;" for convenience, we require $\bar{\mathbf{B}}$ to be orthogonal (i.e., $\bar{\mathbf{B}}\bar{\mathbf{B}}^T = \mathbf{I}_k$, where $\mathbf{I}_k$ is the k-dimensional identity matrix). Since the attacker succeeds only if they can induce a basis different from the true $\mathbf{B}$, we require that $\bar{\mathbf{B}}$ does not span of $\mathbf{X}_\star$, which is equivalent to the condition that $\mathbf{X}_\star\bar{\mathbf{B}}^T\bar{\mathbf{B}} \neq \mathbf{X}_\star$. Thus, this optimization problem seeks $n - n_1$ rows of $\mathbf{X}_\star$, where $\mathcal{I}$ is the corresponding index set. The objective is to minimize the distance between $\mathbf{X}_0^{\mathcal{I}}$ and the span space of the target basis $\bar{\mathbf{B}}$, (i.e., $||\mathbf{X}_0^{\mathcal{I}} - \mathbf{U}\bar{\mathbf{B}}||$).

Algorithm 7.16 Exact Basis Recovery Algorithm

Solve the following optimization problem and get $\mathcal{I}$.

$$\min_{\mathcal{I},\mathbf{L}} ||\mathbf{X}^{\mathcal{I}} - \mathbf{L}||$$
$$\text{s.t. } \text{rank}(\mathbf{L}) \leq k, \mathcal{I} \subseteq \{1, ..., n + n_1\}, |\mathcal{I}| = n \tag{7.8}$$

return a basis of $\mathbf{X}^{\mathcal{I}}$.

To understand the importance of $SR(\mathbf{X}_0)$, consider Algorithm 7.16 for recovering the basis $\mathbf{B}$ of $\mathbf{X}_\star$. If the optimal objective value of optimization problem (7.7), $SR(\mathbf{X}_0)$, exceeds the noise $NR(\mathbf{X}_0)$, it follows that the defender can obtain the correct basis $\mathbf{B}$ using Algorithm 7.16, as it yields a better low-rank approximation of $\mathbf{X}$ than any other basis. Else, it is, indeed, possible for the adversary to induce an incorrect choice of a basis. The following theorem formalizes this argument.

Theorem 7.7   *If $SR(\mathbf{X}_0) \leq NR(\mathbf{X}_0)$, then no algorithm can recover the exact subspace of $\mathbf{X}_\star$ with probability $> 1/2$. If $SR(\mathbf{X}_0) > NR(\mathbf{X}_0)$, then Algorithm 7.16 recovers the true basis.*

7.4.3   EFFICIENT ROBUST SUBSPACE RECOVERY

Consider the objective function (7.8). Since $\text{rank}(\mathbf{L}) \leq k$, we can rewrite $\mathbf{L} = \mathbf{UB}^T$ where $\mathbf{U}$'s and $\mathbf{B}$'s shapes are $n \times k$ and $m \times k$, respectively. Therefore, we can rewrite objective (7.8) as

$$\min_{\mathcal{I},\mathbf{U},\mathbf{B}} ||\mathbf{X}^{\mathcal{I}} - \mathbf{UB}^T|| \ \text{ s.t. } \ |\mathcal{I}| = n \tag{7.9}$$

which is equivalent to

$$\min_{\mathbf{U},\mathbf{B}} \sum_{j=1}^{n} \{||x_i - u_i\mathbf{B}^T|| \text{ for } i = 1, ..., n + n_1\}_{(j)}, \tag{7.10}$$

where x_i and u_i denote the ith row of $\mathbf{X}$ and $\mathbf{U}$ respectively. We can solve Problem 7.10 using alternating minimization, which iteratively optimizes the objective for $\mathbf{U}$ and $\mathbf{B}$ while fixing the other. Specifically, in the tth iteration, we optimize for the following two objectives:

$$\mathbf{U}^{t+1} = \text{argmin}_U ||\mathbf{X} - U(\mathbf{B}^w)^T|| \tag{7.11}$$

$$\mathbf{B}^{t+1} = \text{argmin}_B \sum_{j=1}^{n} \{||x_i - u_i^{w+1}B^T|| \text{ for } i = 1, ..., n + n_1\}_{(j)}. \tag{7.12}$$

Notice that the second step computes the entire $\mathbf{U}$ regardless of the sub-matrix restriction. This is because we need the entire $\mathbf{U}$ to be computed to update $\mathbf{B}$. The key challenge is to compute $\mathbf{B}^{t+1}$ in each iteration, which is, again, a trimmed optimization problem.

7.5   AN EFFICIENT ALGORITHM FOR TRIMMED OPTIMIZATION PROBLEMS

As illustrated above, an important tool for solving robust learning problems such as robust regression and robust subspace recovery is the *trimmed optimization problem* of the form

$$\min_{w} \sum_{j=1}^{n} \{l(y_i, f_w(x_i)) \text{ for } i = 1, ..., n + n_1\}_{(j)}, \tag{7.13}$$

where $f_w(x_i)$ computes the prediction over x_i using parameters w, and $l(\cdot, \cdot)$ is the loss function. It can be shown that solving this problem is equivalent to solving

$$\min_{w,\tau_1,...,\tau_{n+n_1}} \sum_{i=1}^{n+n_1} \tau_i l(y_i, f_w(x_i))$$
$$\text{s.t. } 0 \leq \tau_i \leq 1, \sum_{i=1}^{n+n_1} \tau_i = n. \tag{7.14}$$

We can use the alternating minimization technique to solve this problem, by optimizing for w, and τ_i in an alternating fashion. We present this in Algorithm 7.17. In particular, this algorithm

---

Algorithm 7.17 Trimmed Optimization

---

1. Randomly assign $\tau_i \in \{0, 1\}$ for $i = 1, ..., n + n_1$, such that $\sum_{i=1}^{n+n_1} \tau_i = n$;

2. Optimize $w \leftarrow \text{argmin}_w \sum_{i=1}^{n+n_1} \tau_i l(y_i, f_w(x_i))$;

3. Compute rank_i as the rank of $l(y_i, f_w(x))$ in the ascending order;

4. Set $\tau_i \leftarrow 1$ for $\text{rank}_i \leq n$, and $\tau_i \leftarrow 0$ otherwise;

5. Go to 2 if any of τ_i changes;

6. **return** w.

---

iteratively seeks optimal values for w and $\tau_1, ..., \tau_{n+n_1}$, respectively. Optimizing for w is a standard learning problem. When optimizing $\tau_1, ..., \tau_{n+n_1}$, it is easy to see that $\tau_i = 1$ if $l(y_i, f_w(x_i))$ is among the largest n; and $\tau_i = 0$ otherwise. Therefore, optimizing for $\tau_1, ..., \tau_{n+n_1}$ is a simple sorting step. While this algorithm is not guaranteed to converge to a globally optimal, it often performs well in practice [Liu et al., 2017].

7.6   BIBLIOGRAPHIC NOTES

As we stated in the opening sentence of this chapter, the problem of devising learning algorithms which are robust to data corruption has been studied for several decades. Indeed, what is *new* about data poisoning is a fair question. The key difference is largely about perspective. First, the specific question of *how* to inject malicious noise (from an algorithmic perspective), which we tackled in Chapter 6, is of relatively recent interest. But even when it comes to robust learning, there is a difference between the older research and the more recent approaches. Classical methods generally assume that the fraction of training data which is malicious is extremely small, compared, for example, to accuracy of the classifier. More recent approaches attempt to provide algorithms and guarantees that work even when malicious noise fraction α is a non-negligible proportion of the data.

A number of classical models of learning with malicious noise go back to the mid-1980s. The earliest model (to our knowledge) is due to Valiant [1985], and was subsequently thoroughly analyzed by Kearns and Li [1993], whose algorithm we present as the *data sub-sampling* approach; a similar algorithm for a related *nasty noise* model was proposed by Bshoutya et al. [2002].

A number of subsequent efforts show that variations of linear classifiers in which the weight vector is computed as $w = \sum_i p_i y_i x_i$, where p_i is the probability of data point i (according to a *known* instance distribution), y_i its label, and x_i the feature vector, is robust to a small amount of malicious noise [Kalai et al., 2008, Klivans et al., 2009, Servedio, 2003].

Klivans et al. [2009] proposed the rather powerful and elegant outlier removal idea that we describe, while the approach for detecting outliers based on micromodels was proposed by Cretu et al. [2008]. The RONI algorithm due to Barreno et al. [2010] is closely related to micromodels, but also has an interesting connection to trimmed optimization: this approach attempts to detect outliers by making use of prediction error of the learned model, identifying those points as outliers which introduce a high error. However, RONI has an important limitation in that it assumes that the learner has access to an initial collection of good quality data. A more recent approach by Steinhardt et al. [2017], framed as *certifying* robustness to data poisoning, is also largely in the spirit of outlier detection.

The trimmed optimization approach to robust learning is conceptually related to outlier removal (it attempts to remove instances with high empirical loss with respect to a learned model), but combines it with learning into an effectively single-shot procedure. A number of approaches based on trimmed optimization have been developed to address the problem of learning with malicious noise, both for linear regression [Liu et al., 2017, Xu et al., 2009a], as well as linear classification [Feng et al., 2014]. Our discussion presents the approach by Liu et al. [2017], which requires fewer assumptions compared to prior methods.

A somewhat orthogonal idea to the three classes of approaches we focus on, suggested by Demontis et al. [2017b], is to use l_∞-regularized SVM for increased robustness to poisoning attacks. It is interesting that the same idea has been shown to yield robustness to *evasion* attacks in which the attacker's evasion cost is measured by l_1 regularization.

Finally, a number of approaches consider the problem of robust PCA [Liu et al., 2017, Xu et al., 2012, 2013]. We present the approach by Liu et al. [2017], which has shown good empirical performance compared to some of the others.

CHAPTER 8

Attacking and Defending Deep Learning

In recent years, deep learning has made a considerable splash, having shown exceptional effectiveness in applications ranging from computer vision to natural language processing [Goodfellow et al., 2016]. This splash was soon followed by a series of illustrations of fragility of deep neural network models to small *adversarial* changes to inputs. While initially these were seen largely as robustness tests rather than modeling actual attacks, the language of *adversarial* has since often been taken more literally, for example, with explicit connections to security and safety applications.

As the literature on *adversarial deep learning* recently emerged almost independently of the earlier adversarial machine learning research, and is of considerable independent interest, we focus this chapter solely on attacks on, and defenses of, deep learning. Nevertheless, the content of this chapter is a *special case of decision-time attacks*, and associated defenses—in other words, one must see this chapter as intimately tied to our discussion in Chapters 4 and 5. While there have also been several approaches to poisoning attacks on deep learning, this literature is somewhat less mature at the time of this writing, and we only remark on it in the bibliographic notes to Chapter 6. Since vision applications have been the most important in the adversarial deep learning literature, this chapter is framed in the context of such applications (which are also, conveniently, easiest to visualize).

After a brief description of typical deep learning models, we discuss (decision-time) attacks on deep neural networks. First, we frame such attacks, or *adversarial examples*, as general formal optimization problems. We then discuss three major classes of attacks, which we categorize in terms of the measure of distance they use to quantify the cost of a perturbation.

1. l_2-**norm attacks**: in these attacks the attacker aims to minimize squared error between the adversarial and original image. These typically result in a very small amount of noise added to the image.

2. l_∞-**norm attacks**: this is perhaps the simplest class of attacks which aim to limit or minimize the amount that *any pixel* is perturbed in order to achieve an adversary's goal.

3. l_0-**norm attacks**: these attacks minimize the number of modified pixels in the image.

After discussing the general attack methods in the context of digital images, we describe approaches for making such attacks practical *in the physical world*: that is, adversarially modifying physical objects which are misclassified after being processed into digital form.

Along with the attacks on deep learning, there emerged a number of approaches to mitigate against these. We describe three general approaches for protecting deep learning against decision-time attacks.

1. **Robust optimization**: this is the most theoretically grounded approach, as it aims to directly embed robustness into learning. As such, robust optimization in principle allows one to guarantee or *certify* robustness.

2. **Retraining**: the iterative retraining approach we discussed in Chapter 5 for decision-time attacks can be directly applied to increase robustness of deep learning.

3. **Distillation**: this is a heuristic approach for making gradient-based attacks more difficult to execute by effectively rescaling the output function to ensure that gradients become unstable. It is worth noting that distillation can be defeated by state of the art attacks.

8.1 NEURAL NETWORK MODELS

Deep learning makes use of neural network learning models, which transform an input feature vector through a series of non-linear transformations, called *layers*, before producing a final answer, which for classification is a probability distribution over the classes, while for regression corresponds to real-valued predictions. What makes deep learning *deep* is the fact that one uses many such non-linear transformation layers, where different layers may compute different kinds of functions.

Formally, a deep neural network $F(x)$ over a feature vector x is a composition

$$F(x) = F_n \circ F_{n-1} \circ \cdots \circ F_1(x), \tag{8.1}$$

where each layer $F_l(z_{l-1})$ maps an output z_{l-1} from the previous layer F_{l-1} into a vector z_l as

$$z_l = F_l(z_{l-1}) = g(W_l z_{l-1} + b_l), \tag{8.2}$$

with W_l and b_l the weight matrix and bias vector parameters, respectively, and $g(\cdot)$ a non-linear function, such as a (componentwise) sigmoid $g(a) = 1/(1 + e^a)$ or a rectified linear unit (ReLU), $g(a) = \max(0, a)$. To simplify notation, henceforth we aggregate all of the parameters W_l and b_l of the deep neural network model into a vector θ. We also often omit the explicit dependence of $F(x)$ on θ unless it's necessary for exposition.

In classification settings, the final output of the neural network is a probability distribution p over the classes, i.e., $p_i \geq 0$ and $\sum_i p_i = 1$ for all classes i. This is typically accomplished by having the final layer be a softmax function. To be precise, let $Z(x)$ be the output of the penultimate layer, which is a real number $Z_i(x)$ for each class i. The final output for each class is

$$F_i(x) = p_i = \text{softmax}(Z(x))_i = \frac{e^{Z_i(x)}}{\sum_j e^{Z_j(x)}}. \tag{8.3}$$

A schematic structure of the deep neural network, which emphasizes this relationship, is shown in Figure 8.1. Finally, the predicted class $f(x)$ is the one with the largest probability p_i, i.e.,

$$f(x) = \arg\max_i F_i(x). \tag{8.4}$$

In adversarial settings, one either considers the probabilistic output of the network $F(x)$, or the layer immediately below, $Z(x)$, which has the corresponding real-valued outputs for each class before they are squashed into a valid probability distribution by the softmax function.

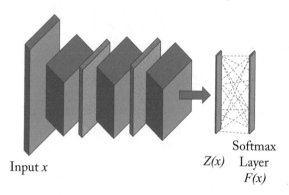

Input x $Z(x)$ Softmax Layer $F(x)$

Figure 8.1: A schematic representation of a deep neural network.

8.2 ATTACKS ON DEEP NEURAL NETWORKS: ADVERSARIAL EXAMPLES

We begin by considering *white-box* attacks, that is, attacks which assume full knowledge of the deep learning model. We subsequently briefly discuss *black-box* attacks.

In typical attacks on deep neural networks, one begins with an original clean image, x_0, and adds noise η to it in order to cause miscategorization, resulting in the adversarially corrupted image x', commonly known as an *adversarial example*. Clearly, adding sufficient noise will always effect a classification error. Consequently, one would either impose a constraint that η is small, formalized by a norm-constraint of the form $\|\eta\| \leq \epsilon$ for some exogenously specified ϵ, or minimize the norm of η. The most common norms which quantify the amount of noise the attacker may add are l_2 (squared error), l_∞ (max-norm), and l_0 (number of pixels modified). We discuss the attacks based on these norms below.

There have been several formulations for the attacker optimization problem described in the literature. The first of these, due to Szegedy et al. [2013], aims to minimize the norm of the added adversarial noise η subject to the constraint that the image is misclassified into a target (and incorrect) class y_T:

$$\min_{\eta} \|\eta\|$$
$$\text{s.t.:} \quad f(x_0 + \eta) = y_T, \quad x_0 + \eta \in [0, 1]^n, \tag{8.5}$$

where the second *box* constraint simply imposes a natural restriction that attacks generate a valid image (with pixels normalized to be between 0 and 1). In our nomenclature, this is a targeted attack. The corresponding reliability attack replaces the first constraint with $f(x_0 + \eta) \neq y$, that is, now the adversary attempts to cause misclassification as *any* class other than the correct label y.

In an alternative version of a reliability attack, proposed by Goodfellow et al. [2015], the attacker's objective is to maximize loss for the constructed image $x' = x_0 + \eta$ with respect to the true label y assigned to x_0:

$$\max_{\eta: \|\eta\| \leq \epsilon} l(F(x_0 + \eta), y). \tag{8.6}$$

Alternatively, the attacker may consider a targeted attack with a target class y_T within the same framework, yielding the following optimization problem:

$$\min_{\eta: \|\eta\| \leq \epsilon} \eta l(F(x_0 + \eta), y_T). \tag{8.7}$$

Perhaps the most important distinction between the different attack approaches proposed in the context of deep learning is the norm that is used to measure the magnitude of adversarial perturbations. Next, we discuss major classes of attacks for each of three norms: l_2, l_∞, and l_0.

8.2.1 l_2-NORM ATTACKS

Attacks in which an attacker is minimizing a Euclidean (l_2) norm of the perturbation either to cause misclassification as a target class (targeted attacks) or simply to cause an error (reliability attack) are among the most potent in practice, and in some cases used as the core machinery for optimizing with respect to other norms. We therefore begin our discussion of attacks against deep learning with this class.

The earliest example of an l_2 attack on deep learning was a targeted attack proposed by Szegedy et al. [2013]. The idea behind this attack is to replace the difficult-to-solve problem (8.5) with a proxy, using the (squared) l_2 norm to quantify the error introduced by the attack:

$$\min_{\eta} c\|\eta\|_2^2 + l(F(x_0 + \eta), y_T)$$
$$\text{s.t.:} \quad x_0 + \eta \in [0, 1]^n. \tag{8.8}$$

The resulting box-contrained optimization problem can be solved using a host of standard techniques for unconstrained optimization (with projection into the box constraint). We can further optimize the coefficient c using line search to find an adversarial example with the smallest l_2 norm.

A more recent targeted l_2 attack was proposed by Carlini and Wagner [2017] (henceforth, the CW attack), who improve the optimization algorithm by using a better objective function.

The starting point for the CW attack is, again, the optimization problem (8.5). Their first step is to reformulate the challenging constraint $f(x_0 + \eta) = y_T$. This is done by constructing a function $h(x_0 + \eta; y_T)$ such that $h(x_0 + \eta; y_T) \leq 0$ iff $f(x_0 + \eta) = y_T$. They consider several candidates for $h(\cdot)$; the one which yields the best performance is

$$h(x_0 + \eta; y_T) = \max\{0, \max_{j \neq y_T} Z(x_0 + \eta)_j - Z(x_0 + \eta)_{y_T}\}. \tag{8.9}$$

As it turns out, using the raw $Z(x)$ rather than the softmax-filtered probability distribution $F(x)$—or the loss function with $F(x)$ as an argument—makes the attack significantly more robust to some of the defensive approaches, such as the distillation defense which we discuss below.

The next step in the CW attack is to reformulate the modified constrained optimization problem by moving the constraint into the objective, analogously to what was done by Szegedy et al., obtaining:

$$\min_{\eta} \|\eta\|_p^p + c \cdot h(x_0 + \eta, y_T) \tag{8.10}$$

$$\text{s.t.:} \quad x_0 + \eta \in [0, 1]^n. \tag{8.11}$$

The l_2 attack then involves the standard gradient descent approach. The simplest way to handle the box constraint is to simply include a projection method in gradient descent which clips any intermediate image to the $[0, 1]$ interval. It turns out that this attack with the objective function above tends to have the best performance.[1] Finally, the parameter of the objective c is chosen as the smallest such parameter which ensures that the attack is successful. A visual illustration of the CW l_2 attack is shown in Figure 8.2.

Another technique for l_2-norm-based attacks, termed *DeepFool*, leverages a linear approximation of a Neural Network [Moosavi-Dezfooli et al., 2016a]. In contrast to the attacks by Szegedy et al. [2013] and Carlini and Wagner [2017], DeepFool implements a *reliability* attack.

To understand the DeepFool attack, we start by assuming that the classifier $F(x)$ is linear, i.e., $F(x) = Wx + b$, and $f(x) = \arg\max_i F_i(x)$ as before. In this case, the optimal attack is a

---

[1]Interestingly, in the original paper the main formulation of the attack involved another option which uses a change of variables to eliminate the need for the box constraint. However, the results in the paper suggest that the simple projected gradient descent tends to perform as well as, or better than, this variant for the specific objective function we discuss here.

Figure 8.2: Illustration of the CW l_2 attack. Left: original image (classified correctly as a jeep). Middle: (magnified) adversarial noise. Right: perturbed image (misclassified as a minivan).

solution of the optimization problem

$$\min_{\eta} \|\eta\|_2^2 \tag{8.12}$$
$$\text{s.t.:} \quad \exists k : w_k^T (x_0 + \eta) + b_k \geq w_{f(x_0)}^T (x_0 + \eta) + b_{f(x_0)},$$

where w_k is the kth row of W corresponding to the weight vector of $F_k(x) = w_k x + b_k$. We can characterize the optimal solution to this problem in closed form. Note that the attacker succeeds if there is some k such that $F_k(x_0 + \eta) - F_{f(x_0)}(x_0 + \eta) \geq 0$ (we break the tie here in the attacker's favor). Define $\tilde{F}_k(x) = F_k(x) - F_{f(x_0)}(x)$. If we further define a hyperplane corresponding to $\tilde{F}_k(x) = 0$, the shortest distance to this hyperplane from x_0 is

$$\delta_k = \frac{|\tilde{F}_k(x_0)|}{\|w_k - w_{f(x_0)}\|_2}, \tag{8.13}$$

and the corresponding optimal η_k (moving δ_k in the orthogonal unit direction toward the hyperplane, $\frac{w_k - w_{f(x_0)}}{\|w_k - w_{f(x_0)}\|_2}$) is then

$$\eta_k = \frac{|\tilde{F}_k(x_0)|}{\|w_k - w_{f(x_0)}\|_2^2}(w_k - w_{f(x_0)}). \tag{8.14}$$

Since the goal of the attacker is to move to the nearest $\tilde{F}_k(x) = 0$ over all k (which is the easiest way to get misclassified as some class other than $f(x_0)$), the optimal solution to Problem 8.12 is then to first choose the closest alternative class

$$k^* = \arg\min_k \delta_k,$$

and then set $\eta^* = \eta_{k*}$, the optimal displacement vector η_k for the class k^*.

The ideas above, of course, do not immediately generalize to non-linear classifiers, such as deep neural networks. However, DeepFool makes use of them in an iterative procedure which repeatedly approximates each $F_k(x)$ by a linear function using a Taylor approximation:

$$F_k(x; x_t) \approx F_k(x_t) + \nabla F_k(x_t)^T x. \tag{8.15}$$

In this approximation, then, $b_k = F_k(x_t)$ and $w_k = \nabla F_k(x_t)$ for a particular x_t obtained in prior iteration t. The next iterate x_{t+1} is then computed as

$$x_{t+1} = x_t + \eta_t, \qquad (8.16)$$

where η_t is the optimal solution as described above for the linear approximation of the neural network function around the previously found x_t. The iterative procedure terminates once x_t with $f(x_t) \neq f(x_0)$ is found, and returns $\eta = \sum_t \eta_t$.

8.2.2 l_∞-NORM ATTACKS

One of the earliest approaches for reliability attacks on deep learning was proposed by Goodfellow et al. [2015], who termed it the *fast gradient sign method (FGSM)*. The goal of Goodfellow et al. [2015] is to approximately solve problem (8.6) with the max-norm constraint. In other words, the attacker's goal is to induce prediction error by adding arbitrary noise η to an original clean image x_0 with the constraint that $\|\eta\|_\infty \leq \epsilon$.

While the loss maximization problem (8.6) is difficult to solve exactly, the key idea in FGSM is to linearize the loss around (x_0, y), where y is the correct label, obtaining

$$\tilde{l}(\eta) = l(F(x_0), y) + \nabla_x l(F(x_0), y)\eta. \qquad (8.17)$$

The optimal solution to the linearized version is then to maximally distort by ϵ independently along each coordinate. This is done in the direction of the sign of the loss gradient for a reliability attack:

$$\eta^* = \epsilon \, \text{sgn}(\nabla_x l(F(x_0), y)). \qquad (8.18)$$

The approach is also easy to apply in the context of a targeted attack: in this case, it is optimal to distort each pixel in the opposite direction of loss with respect to a target class y_T (corresponding to gradient descent, rather than ascent):

$$\eta^* = -\epsilon \, \text{sgn}(\nabla_x l(F(x_0), y_T)). \qquad (8.19)$$

A visual illustration of the FGSM attack is shown in Figure 8.3.

It's worth remarking that the FGSM attack is actually a special case of a more general class of attacks where we impose a constraint $\|\eta\|_p \leq \epsilon$ for an arbitrary p [Lyu et al., 2015]. In this case, the optimal solution for η generalizes to

$$\eta^* = \epsilon \, \text{sgn}(\nabla_x l(F(x_0), y)) \left(\frac{|\nabla_x l(F(x_0), y))|}{\|\nabla_x l(F(x_0), y))\|_q} \right), \qquad (8.20)$$

where l_q is the dual norm of l_p, i.e.,

$$\frac{1}{p} + \frac{1}{q} = 1.$$

Figure 8.3: Illustration of the FGSM attack with $\epsilon = 0.004$. Left: original image (classified correctly as a jeep). Middle: (magnified) adversarial noise. Right: perturbed image (misclassified as a minivan). Note that the added noise is substantially more perceptible for this attack than for the CW l_2 attack in Figure 8.2.

The FGSM attack is a single-step gradient update. This is extremely efficient, but also limits the power of the attacker. A significantly more powerful idea is to perform what is effectively iterative *trust region optimization* [Conn et al., 1987], where we iteratively linearize the objective, optimize the resulting objective within a small *trust region* around the current estimate, and then update both the estimate and the trust region and repeat.

Formally, let β_t be an update parameter (or the trust region around a current estimate with respect to an l_∞ norm). Let x_t be the modified adversarial image in iteration t (starting with x_0 in iteration 0). Then for a reliability attack

$$x_{t+1} = \text{Proj}_\epsilon[x_t + \beta_t \, \text{sgn}(\nabla_x l(F(x_t), y))], \tag{8.21}$$

where Proj_ϵ projects its argument into the feasible space where $\|x_0 - x_{t+1}\|_\infty \leq \epsilon$, which can be done simply by clipping off any single-dimension modifications which exceed ϵ. The corresponding variation for targeted attacks is immediate, with the "+" sign replaced by a "−" sign. A visual illustration of this attack, which has come to be known as the *projected gradient descent (PGD)* attack [Madry et al., 2018, Raghunathan et al., 2018, Wong and Kolter, 2018], is shown in Figure 8.4.

Figure 8.4: Illustration of the iterative GSM attack, which uses eight gradient steps. Left: original image (classified correctly as a jeep). Middle: (magnified) adversarial noise. Right: perturbed image (misclassified as a minivan).

Another variant of a l_∞ attack was proposed by Carlini and Wagner [2017]. Carlini and Wagner first replace the $\|\eta\|_\infty$ term with a proxy

$$\sum_i \max\{0, \eta_i - \tau\}, \tag{8.22}$$

where τ is an exogenously specified constant, which starts at 1 and is decreased in each iteration. Then, the problem is iteratively solved, where if $\eta_i \leq \tau$ in a given iteration for every i (i.e., the cost term of the optimization problem is 0), τ is decreased by a factor of 0.9, and the process repeats.

8.2.3 l_0-NORM ATTACKS

The final set of attack approaches we discuss limit the number of pixels an attacker modifies. The first, *Jacobian-based Saliency Map Attack (JSMA)* attack (a variation of the attack introduced by Papernot et al. [2016b]) aims to minimize the number of modified pixels in an image to cause misclassification as a particular target class y_T—that is, it is a targeted l_0 norm attack. The attack starts with the original image x_0, and then greedily modifies pairs of pixels at a time. The choice of the pair i, j to change is guided by a heuristic based on two quantities:

$$\alpha_{ij} = \frac{\partial Z_{y_T}(x_0)}{\partial x_i} + \frac{\partial Z_{y_T}(x_0)}{\partial x_j}, \tag{8.23}$$

and

$$\beta_{ij} = \sum_k \left(\frac{\partial Z_k(x_0)}{\partial x_i} + \frac{\partial Z_k(x_0)}{\partial x_j} \right) - \alpha_{ij}, \tag{8.24}$$

where y_T is the target class for the attack, as above. Thus, α_{ij} indicates the impact of changing pixels i and j on the target class y_T, while β_{ij} is the indicator of the impact of changing these pixels on other classes. Since y_T is our target, we wish to make α_{ij} as large as possible, while making β_{ij} small. Each pair (i, j) is then assigned a saliency score

$$s_{ij} = \begin{cases} 0 & \alpha_{ij} < 0 \quad \text{or} \quad \beta_{ij} > 0 \\ -\alpha_{ij}\beta_{ij} & \text{o.w.} \end{cases} \tag{8.25}$$

The attack chooses a pair of pixels (i, j) to modify that maximize the saliency s_{ij}. The corresponding pair can be modified in a variety of ways, such as exhaustive search in the discrete space of pixel values, or gradient-based optimization restricted to that pair of pixels.

Carlini and Wagner [2017] propose a different method for implementing an l_0 attack which makes use of their carefully engineering l_2 attack as a subroutine. In their l_0 attack, Carlini and Wagner iteratively apply l_2 attacks to successively shrinking parts of the image. The general idea is to eliminate the pixels which appear to be least critical to attack success in each iteration. In particular, each iteration excludes a pixel i with the smallest value of $\nabla h(x_0 + \eta)_i \eta_i$, where $h(\cdot)$ is the proxy objective function CW also use in their l_2 attack, discussed in Section 8.2.1.

8.2.4   ATTACKS IN THE PHYSICAL WORLD

The attacks described thus far, as most other attacks in the literature on adversarial deep learning, assume that the attacker has direct access to the underlying digital image, which they can modify in arbitrary ways. Realizing such attacks in practice, however, would often involve a modification to the actual physical objects being photographed.

Several efforts have been made both in devising compelling threat models for physical attacks on vision systems, as well as implementing such attacks. We briefly discuss two of these. The first, due to Sharif et al. [2016], developed attacks on face recognition systems (for example, for biometric authentication) which use deep neural networks. In their attack, the attacker wears a printed pair of eyeglass frames which embed the adversarial noise. The second is due to Evtimov et al. [2018], who describe two attacks, one which prints a custom stop sign poster that can be overlaid on the actual stop sign, and another which prints stickers that look like conventional vandalism.

Physical attacks face three additional common challenges to be successfully realizable. First, they must be unobtrusive, a relatively ill-defined concept that aims to capture the likelihood that the attack is discovered before it succeeds. In common adversarial example attacks on deep learning, this is captured by minimizing or limiting the magnitude of the perturbation so that the new image is visually indistinguishable from the original. The two physical attacks above, on the other hand, leverage a more subtle form of psychological hacking, where the attack is hidden in plain sight by virtue of its similarity to common non-adversarial behavior, such as vandalism or wearing glasses. Second, they must account for the ability to physically produce an attack which is optimized in the digital realm. This is captured by having an explicit printability term in the objective. Specifically, both of the attacks above include a non-printability-score term (NPS), defined for a given image x as

$$NPS(x) = \sum_i \prod_{b \in B} |x_i - b|,$$ (8.26)

where i ranges over the pixels in the image and B is a set of printable color configurations in the RGB domain. To further improve the likelihood of success, Sharif et al. suggest constructing a color map which corrects for any discrepancy between digital and printed colors. Third, physical instances, as they present themselves in vision applications in practice, admit a variation of actual object positions in the image, such as slight rotation. A successful attack must be robust to such variations. Both approaches discussed above tackle this problem by using a collection of images X for the same target object, such that a single perturbation η is effective on all, or most of them. For example, the variant of the targeted attack discussed by Sharif et al. modifies Eq. (8.7) into

$$\min_\eta \sum_{x_0 \in X} l(F(x_0 + \eta), y_T).$$ (8.27)

8.2.5   BLACK-BOX ATTACKS

In Chapter 4 we discussed black-box decision-time attacks both in theoretical terms, and in practical implications. Similar ideas largely apply in the context of attacks on deep neural networks. For example, query-based attacks have been explored and found effective by a number of efforts [Bhagoji et al., 2017, Papernot et al., 2016c, 2017]. Papernot et al. [2016c] and Szegedy et al. [2013] also observed a phenomenon that has come to be called *transferability*: adversarial examples designed against one learned model often work against another model that was learned for the same problem (this was observed in the context of deep learning, as well as for other learning paradigms [Papernot et al., 2016c]). The specific illustrations of such transferability are: (a) when models have been trained on different, but related datasets (what we have called *proxy data* in Chapter 4); (b) when a proxy model has been learned based on data collected from queries to the target model (the problem addressed conceptually in Theorem 4.4); and (c) when a different algorithm (proxy algorithm) is used than the one that produces the target model (for example, training and attacking a deep neural network to generate adversarial examples against a logistic regression model).

8.3   MAKING DEEP LEARNING ROBUST TO ADVERSARIAL EXAMPLES

Observations of vulnerabilities in deep neural networks naturally spawned a literature attempting to defend deep learning against attacks. We now describe some of the prominent proposals for robust deep learning in adversarial settings. Throughout this section, we will now explicitly represent the dependence of the deep learning model on its parameters θ, using notation $F(x, \theta)$, to capture the fact that we are at this point attempting to *learn* the parameters θ which lead to adversarially robust predictions.

8.3.1   ROBUST OPTIMIZATION

Recall from Chapter 5 that the problem of robust learning amounts to the problem of adversarial risk minimization, where the objective function captures modifications to feature vectors made by an adversary. Let $\mathcal{A}(x, \theta)$ be the adversarial model which returns a feature vector x'—an adversarial example—given an input x and a collection of (deep neural network) model parameters θ. If we assume that every possible instance in our training data $\mathcal{D}$ may give rise to adversarial behavior, the adversarial empirical risk minimization problem becomes

$$\sum_{i \in \mathcal{D}} l(F(\mathcal{A}(x_i; \theta), \theta), y_i). \tag{8.28}$$

Applying a zero-sum game relaxation (upper bound) of Equation (8.28), we obtain a *robust optimization* formulation of the problem of learning an adversarially robust deep neural network:

$$\min_{\theta} \sum_{i \in \mathcal{D}} \max_{\eta:\|\eta\|\leq\epsilon} l(F(x_i + \eta, \theta), y_i) \tag{8.29}$$

if we constrain attacks to modify an original image at most ϵ is some target norm. This robust optimization framework for reasoning about adversarially robust deep learning proved extremely fruitful. In particular, it has given rise to three general ideas for robust deep learning.

1. **Adversarial Regularization**: (also known as *adversarial training*) uses an approximation of the worst-case loss function with respect to small changes in the image as a regularizer.

2. **Robust Gradient Descent**: uses the gradient of the worst-case loss function in learning a robust deep neural network.

3. **Certified Robustness**: uses the gradient of a convex upper bound on the worst-case loss function to learn a robust deep network.

Next, we briefly describe these three ideas. For convenience, we define the worst-case loss as

$$l_{wc}(F(x, \theta), y) = \max_{\eta:\|\eta\|\leq\epsilon} l(F(x + \eta, \theta), y). \tag{8.30}$$

Adversarial Regularization

The issue with the robust optimization approach above is that it may be too conservative, since in most settings adversaries are not actively manipulating images. A common generalization is to explicitly trade off between accuracy on original non-adversarial data, and robustness to adversarial examples:

$$\min_{\theta} \sum_{i \in D} \left[\alpha l(F(x_i; \theta), y_i) + (1 - \alpha) l_{wc}(F(x_i, \theta), y_i) \right], \tag{8.31}$$

where α is the exogenous parameter trading off these two considerations. We can then think of the term $l_{wc}(F(x, \theta), y)$ as adversarial regularization.

Typically, adversarial regularization (training) is only practical with a simplified attack model, such as the gradient-sign attacks. In particular, consider the generalization of FGSM in Section 8.2.2 where the attacker faces the constraint $\|\eta\|_p \leq \epsilon$ for a general l_p norm. In this case, by applying Taylor approximation as in the associated attacks, the adversarial regularization problem can be transformed into

$$\min_{\theta} \sum_{i \in D} \left[\alpha l(F(x_i; \theta), y_i) + (1 - \alpha) \|\nabla_x l(F(x_i; \theta), y_i)\|_q \right], \tag{8.32}$$

where l_q is the dual norm of l_p and we simply plug in the optimal attack η^* from Equation (8.20).

Equation (8.32) shows that regularization based on robust optimization ideas has an interesting structure: it amounts to regularizing based on the magnitude of the gradient of the loss function. This has a certain intuitive appeal: very large gradients imply a more unstable model under small perturbations (since small changes in x can lead to large changes in the output), and regularizing these should improve robustness to adversarial examples. On the other hand, this regularization is somewhat unusual, as it depends on the specific instances x, whereas regularization terms typically only depend on the model parameters θ.

Robust Gradient Descent

Adversarial regularization ultimately involves a heuristic way to approximate the robust optimization problem faced by the learner. Stepping back, we may note that in principle it would suffice to have a way to calculate gradients of the worst-case loss function, $l_{wc}(F(x, \theta), y)$, to enable training a robust deep neural network using standard stochastic gradient descent methods. However, this loss function is not everywhere differentiable. Moreover, the optimization problem involved in computing the worst-case loss itself is intractable, suggesting that computing its gradient may also be computationally challenging.

Madry et al. [2018] observe, however, that an application of an important result from robust optimization allows us to compute gradients approximately, and use these in a gradient descent training procedure in practice. In particular, Madry et al. [2018] derive the following as a corollary of Danskin's classic result in robust optimization [Danskin, 1967].

Proposition 8.1 Madry et al.   *Let (x, y) be arbitrary feature vector and label pair, and suppose that η^* is a maximizer of $\max_{\eta \mid \|\eta\|_p \leq \epsilon} l(F(x + \eta, \theta), y)$. Then, as long as it's non-zero, $-\nabla_\theta l(F(x + \eta^*, \theta), y)$ is a descent direction for $l_{wc}(F(x, \theta), y)$.*

The upshot of Proposition 8.1 is that during stochastic gradient descent, when we are considering a datapoint (x, y) and have a current parameter estimate θ, we can use the gradient of the loss function at any optimal solution for worst-case loss η^* to take the next gradient descent step. Significantly, Madry et al. observe that in practice even good quality approximate optimizers of worst-case loss, such as PGD in Section 8.2.2, appear to suffice for training a robust neural network using this approach.[2]

Certified Robustness

Robust gradient descent is a principled advance over the more simplistic adversarial regularization approach, but it is still heuristic, albeit with strong empirical support, and cannot *guarantee* robustness. Several approaches recently emerged for both deriving *certificates*, or guarantees, of robustness to adversarial perturbations in a particular class, and training neural networks which

---

[2]It is worth noting that the application of this approach in stochastic gradient descent is essentially a special case of the stochastic gradient descent variant of retraining; see Li and Vorobeychik [2018] for a brief discussion of the latter.

minimize a provable upper bound on adversarial risk [Raghunathan et al., 2018, Wong and Kolter, 2018].

The key idea behind certified robustness is to obtain a tractable upper bound $J(x, y, \theta)$ on the worst-case loss:

$$l_{wc}(F(x, \theta), y) \leq J(x, y, \theta).$$

Both Raghunathan et al. [2018] and Wong and Kolter [2018] do this in two steps: (1) they obtain a convex relaxation of the optimization problem for computing $l_{wc}(F(x, \theta), y)$; and then (2) use a dual of this convex optimization problem. They key insight in both approaches is that *any feasible solution* of the dual yields an upper bound on the primal, which in turn is an upper bound on worst-case loss, and both choose a particular feasible solution.

We illustrate some aspects of this general approach based on Wong and Kolter [2018], who assume that the neural networks use ReLU activation functions. Their first step is to relax ReLU activations $b = \max\{0, a\}$ using a collection of linear inequalities:

$$b \geq 0, b \geq a, -ua + (u - l)b \geq -ul,$$

where u and l are upper and lower bounds on the activation values, respectively. With this relaxation, computation of the upper bound on worst-case loss can be represented as a linear program, albeit with many variables. By strong duality of linear programming, the dual solution of this linear program is also an upper bound on the worst-case loss. Moreover, any feasible solution still produces an upper bound. Thus, by fixing the values of a subset of dual variables, Wong and Kolter [2018] devised a linear algorithm to compute an upper bound $J(x, y, \theta)$ on the worst-case loss. Significantly, this upper bound is differentiable with respect to θ, and can be used as a part of the stochastic gradient descent approach for training a robust neural network. The full algorithm developed by Wong and Kolter [2018] (including the derivation of upper and lower bounds for ReLU activation units) is quite involved; we defer the readers to the original paper for technical details.

8.3.2   RETRAINING

As we already mentioned, attacks on deep learning described in this chapter are special cases of what we have called *decision-time attacks*. As such, the general-purpose defense by iterative retraining described in Section 5.3.2 applies directly. In particular, we can iteratively train the neural network, attack it by generating adversarial examples according to any of the attack models described, adding these to training data, and repeating the process. An important advantage of this iterative retraining approach is that it is agnostic as to which algorithm is used to construct adversarial examples. In contrast, the approaches based on robust optimization all effectively assume reliability attacks, which may result in solutions (or certificates of robustness) that are too conservative in practice.

8.3.3 DISTILLATION

Distillation is a heuristic technique for training deep neural networks initially proposed for transferring knowledge from a more to a less complex model (essentially, for compression). Papernot et al. [2016a] proposed using distillation to make deep neural networks more robust to adversarial noise.

The distillation approach works as follows.

1. Start with the original training dataset $D = \{x_i, y_i\}$, where labels y_i are encoded as one-hot vectors (i.e., all zeros, except for a 1 in the position corresponding to the true class of x_i).

2. Train a deep neural network after replacing the softmax function in the softmax layer with

$$F_i(x) = \frac{e^{Z_i(x)/T}}{\sum_j e^{Z_j(x)/T}}$$

   for an exogenously chosen shared *temperature* parameter T.

3. Create a new training dataset $D' = \{x_i, y_i'\}$, where $y_i' = F(x_i)$ with $F(\cdot)$ the soft (probabilistic) class labels returned by the previously trained neural network.

4. Train a new deep neural network with the same temperature parameter T as the first one, but on the new dataset D'.

5. Use the retrained neural network after eliminating the temperature T from the final (softmax) layer (i.e., setting $T = 1$ at test time, scaling the softmax terms down by a factor of T, and thereby increasing the sharpness of predicted class probabilities).

While defensive distillation was shown to be quite effective against several classes of attacks, such as FGSM and JSMA (however, replacing $Z_i(x)$ with $F_i(x)$ for JSMA in the experiments; see Carlini and Wagner [2017]), the CW attack was later demonstrated to effectively defeat it. The crucial insight offered by Carlini and Wagner is that distillation appears to be effective against originally crafted attacks because the temperature parameter forces the values $Z_i(x)$ to be amplified, and once T is set to 1, this in turn results in extremely sharp class predictions, so much so that gradients become numerically unstable. However, if the attacker uses the last hidden layer values $Z(x)$, gradients again become well-behaved.

8.4 BIBLIOGRAPHIC NOTES

The literature on adversarial examples in deep learning was spawned with the publication of Szegedy et al. [2013]. Their goal was largely to show that despite state-of-the-art performance on benchmark image datasets, deep learning models appear to be very fragile to several forms of "gamesmanship" with the images. For example, they showed that they can design images which are unrecognizable to a human, but are reliably classified as a target class. In addition, they showed the impact of introducing a small amount adversarial noise. A long series of papers followed this initial demonstration, considering many variations of attacks.

As mentioned above, attack approaches fall into three rough categories based on the norm they use to measure the amount of noise introduced into an image. Szegedy et al. [2013] were the first to propose l_2 norm attacks, which were improved successively by Moosavi-Dezfooli et al. [2016b], and then by Carlini and Wagner [2017] (who also developed l_0 and l_∞ norm attacks). Soon after the first work on l_2 norm attacks was published, a simple FGSM method for l_∞ attacks was developed by several of the same authors [Goodfellow et al., 2015]. The idea of FGSM was to use a first-order Taylor expansion to approximate the loss function that the attacker aims to optimize. Subsequently, this idea was extended to other l_p-norm attacks by Lyu et al. [2015].

While most approaches for attacking deep neural networks in the literature are white-box attacks, which assume that the deep learning model is known to the attacker, several efforts demonstrated the phenomenon of transferability of adversarial examples, enabling effective black-box attacks [Papernot et al., 2016c, 2017, Szegedy et al., 2013]. The high-level observation is that frequently attacks on one deep neural network can be effective against others trained to address the same prediction problem.

Two other major issues in the literature on attacking deep neural networks are worth noting: (1) designing adversarial noise which is effective when added to multiple images simultaneously, and (2) attacks in the physical world which cause erroneous predictions after their digital representation is fed into a deep neural network. The first of these was addressed by Moosavi-Dezfooli et al. [2017], who describe an attack in which a single adversarial noise is generated which can be added to all images, and successfully cause misclassification by a state-of-the-art deep neural network. The success of this attack is quite surprising, as one would have previously been skeptical that such universal adversarial perturbations are possible. The second issue has been addressed by several efforts. In one of these that we described above, Sharif et al. [2016] demonstrate that specially designed glass frames can be printed which would defeat authentication approaches based on face recognition, as well as video surveillance techniques. In another, also discussed in this chapter, Evtimov et al. [2018] demonstrate that adversarial pertubations can be robustly implemented in the physical domain in order to fool deep learning classifiers of traffic signs. In another related effort, Kurakin et al. [2016] demonstrate that they can introduce adversarial perturbations even after images are printed and subsequently again digitized.

Defending deep learning against adversarial perturbation attacks amounts to developing techniques for learning more robust deep neural network models. A natural framework within which to consider robust learning is robust optimization, where the learner aims to minimize worst-case loss with respect to arbitrary perturbations within an ϵ ball (measured according to some l_p norm; commonly, this is the l_∞ norm).

The very first paper that discussed an l_2 attack on deep learning, Szegedy et al. [2013], also proposed a simple defensive approach through iterative retraining. The paper which followed and introduced the FGSM attack [Goodfellow et al., 2015] suggested adversarial regularization (they called it adversarial training) as a solution, an idea which was significantly generalized by Lyu et al. [2015]. Soon after, the cybersecurity community picked up the thread, and distillation was suggested as a defense [Papernot et al., 2016a], an idea which was promptly broken by Carlini and Wagner [2017] and is now generally viewed as ineffective. However, this flurry of ideas led to a formalization of robust deep learning as robust optimization by Madry et al. [2018], Raghunathan et al. [2018], and Wong and Kolter [2018]. Interestingly, the connection between adversarially robust learning in the context of decision-time attacks and robust optimization actually predates these efforts by at least a decade. For example, Teo et al. [2007] already considered learning with invariances using precisely the same robust optimization approach, as did Xu et al. [2009b], who showed equivalence between robust learning and regularization in linear support vector machines. In any event, robust optimization turned out to be a very fruitful connection, as it led to two major advances: first, by Madry et al. [2018], who used Danskin's theory to directly apply stochastic gradient descent to the robust learning formulation (using worst-case loss), and then independently by Raghunathan et al. [2018] and Wong and Kolter [2018] who developed distinct relaxation and duality methods for certifying robustness and gradient-based learning of robust models. In particular, Raghunathan et al. [2018] used semi-definite programming as the core tool, albeit restricted for the moment to two-layer neural networks, whereas Wong and Kolter [2018] relied on a convex polyhedral relaxation of the ReLU activation units, allowing them to upper bound the optimal worst-case loss by a solution to a linear program, for arbitrary ReLU-based deep networks. Neither of these two methods are sufficiently scalable at the moment to tackle realistic image datasets, but both approaches significantly advanced the thinking about robust deep learning.

Several recent heuristic methods for defending deep learning against adversarial examples that we did not discuss at length in this chapter involve a form of anomaly detection [Rouhani et al., 2017], and using layer-level nearest neighbor sets to devise a confidence measure for predictions [Papernot and McDaniel, 2018]. Both of these are ultimately taking steps to address an important issue which has not received much attention: determining confidence in predictions based on similarity of new instances to the distribution of training data.

CHAPTER 9

The Road Ahead

This book provided an overview of the field of adversarial machine learning. Important issues were clearly left out, some deliberately to simplify exposition, others perhaps unintentionally. The field has become quite active in recent years in no small measure due to the attention that attacks on deep learning methods have received. While we devote an entire chapter solely to adversarial deep learning, we emphasize that proper understanding of these necessitates a broader look at adversarial learning that the rest of the book provides.

In this last chapter, we briefly consider the road ahead in adversarial machine learning. We'll start with the active research question of robust optimization as a means for robust learning in the context of decision-time attacks (such as adversarial examples).

9.1 BEYOND ROBUST OPTIMIZATION

Robust optimization has become a major principled means for formalizing the problem of robust learning in the presence of decision-time attacks. Without a doubt, this approach has compelling strengths. Perhaps the most significant of these is a guarantee, or certification, of robustness: if we obtain an expected robust error of e, we can guarantee that *no* adversarial manipulation within the considered class will cause our model to err more than e, on average. While solving such robust learning problems at scale is a drawback, one can envision technical advances that would make it practical in at least some realistic domains. Here, we discuss conceptual limitations of the robust optimization approach to the problem.

The main limitation of robust optimization is that it yields very conservative solutions. There are several reasons for this. First, robust learning formulations maximize the learner's loss. However, loss is typically an upper bound on the true learning objective. For example, in binary classification, arguably the ideal loss function is the 0/1 loss, which returns 0 if the predicted class is correct, and 1 otherwise. In practice, however, convex relaxations of the 0/1 loss are used, such as hinge loss. Additionally, tractability often warrants additional upper bounds on the worst-case loss beyond this, so that the final bound is unlikely to tightly capture the true optimization problem the learner is trying to solve. The advantage of the upper bounding idea is that its guarantees are conservative: robustness guarantees for an upper bound directly transfer to robustness guarantees for the original problem. There are two concerns, however: (1) this may cause an unnecessary sacrifice in performance, both on non-adversarial data, and even on adversarial data; and (2) whatever bounds on adversarial risk one thereby obtains may not be especially meaningful in practice if they are not very tight.

In addition to the issues highlighted above, another way in which robust learning is conservative is that the attacker's objective may not be an arbitrary misclassification. Take a stop sign as an example: it is likely far more concerning (and, presumably, adversarially significant) if it were misclassified as a speed limit sign than if it were misclassified as a yield or a "do not enter" sign.

These concerns should not be used to dismiss robust optimization approaches for robust learning. Rather, our goal is to suggest that research needs also to consider alternative principled approaches to the adversarial learning problem. One such approach is to view it as a Stackelberg game, as we discussed in Chapter 5. In this way, we can consider alternative adversarial models, and still aim to produce the best learning approach, albeit perhaps somewhat more tailored to a specific threat model of interest. Robust optimization is a special case, in which this game model is zero-sum (that is, min-max), but it shouldn't be the *only* approach to consider.

The discussion thus far pertained solely to the issue of decision-time attacks, but some of the same considerations are relevant in the context of poisoning attacks as well. Most approaches to date aim to be robust to *arbitrary* modifications of training data (within some budget constraint, of course, such as a bound on the fraction of data that can be poisoned). In practice, this seems overly conservative: if the attacker deliberately poisoned data, they are unlikely to poison an arbitrary subset of data (if they have that kind of access, they may as well poison all data), and if data is collected from multiple untrusted sources, the attacker is unlikely to compromise all of them. Consequently, an important research direction is how to capture structure imposed on how data may be poisoned. An illustration of such structure is when the training data is a combination of multiple data sources, a subset of which may be attacked, recently modeled by Hajaj and Vorobeychik [2018] as an adversarial task assignment problem (and not specific to machine learning). Similarly, data may be obtained from multiple sensors, and a subset of these may be compromised, with data collected from compromised sensors arbitrarily poisoned. In either case, such structure may allow for better algorithmic techniques for practical robust learning, for example, leveraging approaches for detecting malicious data sources [Wang et al., 2014].

9.2   INCOMPLETE INFORMATION

Framing the problem of defending learning against attacks as a game between the learner and the attacker suggests another research direction: modeling incomplete information players may have about one another. There are two natural classes of incomplete information: information that the attacker has about the learning system (which is fully known to the learner), and information the learner has about the attacker, such as the attacker's objective, and what information the attacker actually has (for example, about proxy data available to the attacker). There have been scarcely any attempts to model incomplete information in developing robust learning approaches; essentially the sole example is by Grosshans et al. [2013], who consider uncertainty about the attacker's relative value of different datapoints, but not any of the other aspects. As an illustration of the challenge faced in modeling robust learning as a game of incomplete information (or Bayesian game), consider the issue of information available to the attacker in a black-box attack. Suppose that the attacker is uncertain about what data is used by the learner, and the learner uses the data to decide which features to use. If the attacker then learns something about which features the learner uses, they can in principle use this information to infer something about the training data as well. However, it is not trivial to construct a compelling model of this encounter, let alone solve the resulting game.

9.3   CONFIDENCE IN PREDICTIONS

The issue of deriving confidence in estimates is fundamental in statistics. However, it is somewhat under-explored in machine learning, where predictions tend to be *point predictions*, without associated confidence. To be sure, even common probabilistic predictions, such as those produced by typical deep neural networks, are in essence point predictions, as they do not necessarily reflect the empirical support a particular prediction has. Indeed, Papernot and McDaniel [2018] essentially argue that this is a major reason for the success of adversarial examples, which by their nature take a learned model out of its "comfort zone," so to speak. The approach Papernot and McDaniel propose can be viewed as an example of *transductive conformal prediction (TCP)*, a general approach for deriving confidence in specific predictions, as well as distributions over predictions, based on empirical support [Vovk et al., 2005]. An alternative way to look at confidence is to consider how unusual, or anomalous, a particular input is, given the training data that was used to derive a model.

Of course, assigning confidence to predictions isn't sufficient: one must still assess what the most reasonable prediction should be, or perhaps selectively refuse to make any prediction when a model is sufficiently uncertain about which label should be assigned. These issues are clearly more general than adversarial learning, but are surely an important aspect of the problem, as low-confidence regions of the sample space often present many of the vulnerabilities exhibited in learning.

9.4   RANDOMIZATION

Randomization is an important tool in security settings, and has been a fundamental part of game theoretic modeling of such problems [Tambe, 2011]. Interestingly, there have been relatively few attempts to introduce randomization into adversarial learning paradigms. One of these we discussed at length in Chapter 5, but it is restricted to binary classification over binary feature spaces. The general challenge in introducing randomization into predictions is that, on the one hand, it may degrade performance on non-adversarial data, and on the other hand, one may be able to design robust attacks that defeat all possible realizations of a randomized model.

9.5   MULTIPLE LEARNERS

Both decision-time attacks and poisoning attacks have almost universally assumed that there is a single target learning system. In practice, there are often many target organizations that combine to form an *ecology* of learners, and it is this ecology that is often attacked. For example, spammers typically target the universe of spam filters, and public malware datasets are used by many organizations to develop machine learning approaches for malware detection. Interestingly, surprisingly little research has studied the problem of attacking multiple learners, or how learning agents would jointly choose to learn in the presence of adversarial threat. We mention two exceptions. The first is the work by Stevens and Lowd [2013] who ask the computational complexity question about evading a collection of binary linear classifiers. The second is a recent effort by Tong et al. [2018b], who study the game induced by a collection of learning models and an attacker who deploys a decision-time attack. In the latter work, which can be seen as an example of multi-defender security games [Smith et al., 2017], the game is in terms of parameters w for a linear regression model, and the choice of transformation of the regression input for the attacker aimed at attacking the combination of all models. In formal terms, it is a two-stage multi-leader Stackelberg game, with learners jointly deciding on their model parameters w_i, and the attacker subsequently choosing the best attack. These two efforts are first steps, but many research opportunities remain, perhaps the most significant of which is generalization of these approaches to non-linear models.

9.6   MODELS AND VALIDATION

Our final remarks concern the very general issue of modeling and validation. In adversarial machine learning, as in security more broadly, a central issue is adversarial modeling. Take decision-time attacks as an illustration. Commonly, an adversary is modeled as modifying a collection of features (such as pixels in an image), with either a constraint on the extent to which such modifications are allowed (typically, measured by the l_p distance), or incorporting a modification cost into an attacker's objective. This is clearly very stylized. For example, adversaries would not typically modify images at the level of pixels, but, perhaps, physical objects that would be subsequently captured into images. Moreover, modifications may include slight spatial transformations, which would appear to be large in standard l_p norm measures, but would remain undetectable to a human [Xiao et al., 2018]. As another example, malware writers attempting to evade detection do not directly manipulate features extracted from malware, but *malware code*. The question of scietific validity of standard stylized models of attacks on machine learning is therefore quite natural. To date, the sole attempt to rigorously study this issue was by Tong et al. [2018a], who investigate validity of conventional *feature space* models of adversarial evasion— that is, models in which the adversary is assumed to directly modify malware features. While this work sheds some light on this issue, for example, both by demonstrating that feature space models can poorly capture adversarial behavior, and by extending such models to improve their validity, the problem of validation remains a major research challenge facing adversarial machine learning.

Bibliography

Scott Alfeld, Xiaojin Zhu, and Paul Barford. Data poisoning attacks against autoregressive models. In *AAAI Conference on Artificial Intelligence*, 2016. 42, 43, 51

Scott Alfeld, Xiaojin Zhu, and Paul Barford. Explicit defense actions against test-set attacks. In *AAAI Conference on Artificial Intelligence*, 2017. 75

Martin Anthony and Peter L. Bartlett. *Neural Network Learning: Theoretical Foundations*. Cambridge University Press, 1999. DOI: 10.1017/cbo9780511624216. 9

Martin Anthony and Peter L. Bartlett. *Neural Network Learning: Theoretical Foundations*. Cambridge University Press, 2009. DOI: 10.1017/cbo9780511624216. 16, 51

Marco Barreno, Blaine Nelson, Russell Sears, Anthony D. Joseph, and J. D. Tygar. Can machine learning be secure? In *ACM Asia Conference on Computer and Communications Security*, pages 16–25, 2006. DOI: 10.1145/1128817.1128824. 24, 50

Marco Barreno, Blaine Nelson, Anthony D. Joseph, and J. D. Tygar. The security of machine learning. *Machine Learning*, 81:121–148, 2010. DOI: 10.1007/s10994-010-5188-5. 24, 25, 102, 103, 111

Dimitri P. Bertsekas and John N. Tsitsiklis. *Neuro-Dynamic Programming*. Optimization and Neural Computation, Athena Scientific, 1996. DOI: 10.1007/0-306-48332-7_333. 17

Arjun Nitin Bhagoji, Warren He, Bo Li, and Dawn Song. Exploring the space of black-box attacks on deep neural networks. *Arxiv Preprint, ArXiv:1712.09491*, 2017. 123

Alexy Bhowmick and Shyamanta M. Hazarika. E-mail spam filtering: A review of techniques and trends. *Advances in Electronics, Communication and Computing*, 2018. DOI: 10.1007/978-981-10-4765-7_61. 10

B. Biggio, B. Nelson, and P. Laskov. Support vector machines under adversarial label noise. In *Proc. of the Asian Conference on Machine Learning*, pages 97–112, 2011. 96

Battista Biggio and Fabio Roli. Wild patterns: Ten years after the rise of adversarial machine learning. *ArXiv:1712.03141*, 2018. 23, 25

Battista Biggio, Blaine Nelson, and Pavel Laskov. Poisoning attacks against support vector machines. In *International Conference on Machine Learning*, 2012. 82, 96

Battista Biggio, Igino Corona, Davide Maiorca, Blaine Nelson, Nedim Srndic, Pavel Laskov, Giorgio Giacinto, and Fabio Roli. Evasion attacks against machine learning at test time. In *European Conference on Machine Learning and Knowledge Discovery in Databases*, pages 387–402, 2013. DOI: 10.1007/978-3-642-40994-3_25. 25, 50

Battista Biggio, Samuel Rota Bulo, Ignazio Pillai, Michele Mura, Eyasu Zemene Mequanint, Marcello Pelillo, and Fabio Roli. Poisoning complete-linkage hierarchical clustering. In *Structural, Syntactic, and Statistical Pattern Recognition*, 2014a. DOI: 10.1007/978-3-662-44415-3_5. 85, 97

Battista Biggio, Giorgio Fumera, and Fabio Roli. Security evaluation of pattern classifiers under attack. *IEEE Transactions on Knowledge and Data Engineering*, 26(4):984–996, 2014b. DOI: 10.1109/tkde.2013.57. 50, 85, 97

Christopher M. Bishop. *Pattern Recognition and Machine Learning*. Information Science and Statistics, Springer, 2011. 8, 16, 55

Mariusz Bojarski, Davide Del Testa, Daniel Dworakowski, Bernhard Firner, Beat Flepp, Prasoon Goyal, Lawrence D. Jackel, Mathew Monfort, Urs Muller, Jiakai Zhang, Xin Zhang, Jake Zhao, and Karol Zieba. End to end learning for self-driving cars. *ArXiv:1604.07316*, 2016. 9

Craig Boutilier, Thomas Dean, and Steve Hanks. Decision-theoretic planning: Structural assumptions and computational leverage. *Journal of Artificial Intelligence Research*, 11(1):94, 1999. 17

Craig Boutilier, Richard Dearden, and Moisés Goldszmidt. Stochastic dynamic programming with factored representations. *Artificial Intelligence*, 121(1):49–107, 2000. DOI: 10.1016/s0004-3702(00)00033-3. 17

Michael Brückner and Tobias Scheffer. Stackelberg games for adversarial prediction problems. In *ACM SIGKDD International Conference on Knowledge Discovery and Data Mining*, pages 547–555, 2011. DOI: 10.1145/2020408.2020495. 63, 74

Michael Brückner and Tobias Scheffer. Static prediction games for adversarial learning problems. *Journal of Machine Learning Research*, (13):2617–2654, 2012. 74

Nader H. Bshoutya, Nadav Eironb, and Eyal Kushilevitz. PAC learning with nasty noise. *Theoretical Computer Science*, 288:255–275, 2002. DOI: 10.1016/s0304-3975(01)00403-0. 96, 110

Jian-Feng Cai, Emmanuel Candès, and Zuowei Shen. A singular value thresholding algorithm for matrix completion. *SIAM Journal on Optimization*, 20(4):1956–1982, 2010. DOI: 10.1137/080738970. 12, 91

Emmanuel Candès and Ben Recht. Exact matrix completion via convex optimization. *Foundations of Computational Mathematics*, 9(6):717–772, 2007. DOI: 10.1007/s10208-009-9045-5. 12, 17, 91

N. Carlini and D. Wagner. Towards evaluating the robustness of neural networks. In *IEEE Symposium on Security and Privacy*, pages 39–57, 2017. DOI: 10.1109/sp.2017.49. 39, 51, 117, 121, 127, 128, 129

Gert Cauwenberghs and Tomaso Poggio. Incremental and decremental support vector machine learning. In *Neural Information Processing Systems*, pages 409–415, 2001. 82

D. H. Chau, C. Nachenberg, J. Wilhelm, A. Wright, and C. Faloutsos. Polonium: Tera-scale graph mining and inference for malware detection. In *SIAM International Conference on Data Mining*, 2011. DOI: 10.1137/1.9781611972818.12. 10

Zhilu Chen and Xinming Huang. End-to-end learning for lane keeping of self-driving cars. In *IEEE Intelligent Vehicles Symposium*, 2017. DOI: 10.1109/ivs.2017.7995975. 9

Andrew R. Conn, Nicholas I. M. Gould, and Philippe L. Toint. *Trust-Region Methods*. Society for Industrial and Applied Mathematics, 1987. DOI: 10.1137/1.9780898719857. 120

Gabriela F. Cretu, Angelos Stavrou, Michael E. Locasto, Salvatore J. Stolfo, and Angelos D. Keromytis. Casting out demons: Sanitizing training data for anomaly sensors. In *IEEE Symposium on Security and Privacy*, pages 81–95, 2008. DOI: 10.1109/sp.2008.11. 102, 111

Nilesh Dalvi, Pedro Domingos, Mausam, Sumit Sanghai, and Deepak Verma. Adversarial classification. In *SIGKDD International Conference on Knowledge Discovery and Data Mining*, pages 99–108, 2004. DOI: 10.1145/1014052.1014066. 34, 36, 50, 74

J. M. Danskin. *The Theory of Max-Min and its Application to Weapons Allocation Problems*. Springer, 1967. DOI: 10.1007/978-3-642-46092-0. 125

Ronald De Wolf. A brief introduction to Fourier analysis on the Boolean cube. *Theory of Computing, Graduate Surveys*, 1:1–20, 2008. DOI: 10.4086/toc.gs.2008.001. 72

A. Demontis, M. Melis, B. Biggio, D. Maiorca, D. Arp, K. Rieck, I. Corona, G. Giacinto, and F. Roli. Yes, machine learning can be more secure! A case study on android malware detection. In *IEEE Transactions on Dependable and Secure Computing*, 2017a. DOI: 10.1109/tdsc.2017.2700270. 75

Ambra Demontis, Battista Biggio, Giorgio Fumera, Giorgio Giacinto, and Fabio Roli. Infinity-norm support vector machines against adversarial label contamination. In *Italian Conference on Cybersecurity*, pages 106–115, 2017b. 111

Carl Eckart and Gale Young. The approximation of one matrix by another of lower rank. *Psychometrika*, 1(3):211–218, 1936. DOI: 10.1007/bf02288367. 107

Ivan Evtimov, Kevin Eykholt, Earlence Fernandes, Tadayoshi Kohno, Bo Li, Atul Prakash, Amir Rahmati, and Dawn Song. Robust physical-world attacks on deep learning visual classification. *Conference on Computer Vision and Pattern Recognition*, 2018. 21, 122, 128

Jiashi Feng, Huan Xu, Shie Mannor, and Shuicheng Yan. Robust logistic regression and classification. In *Neural Information Processing Systems*, vol. 1, pages 253–261, 2014. 111

Prahlad Fogla and Wenke Lee. Evading network anomaly detection systems: Formal reasoning and practical techniques. In *ACM Conference on Computer and Communications Security*, pages 59–68, 2006. DOI: 10.1145/1180405.1180414. 50

Prahlad Fogla, Monirul Sharif, Roberto Perdisci, Oleg Kolesnikov, and Wenke Lee. Polymorphic blending attacks. In *USENIX Security Symposium*, 2006. 28, 50

Drew Fudenberg and David K. Levine. *The Theory of Learning in Games*. Economic Learning and Social Evolution, MIT Press, 1998. 74

Rainer Gemulla, Erik Nijkamp, Peter J. Haas, and Yannis Sismanis. Large-scale matrix factorization with distributed stochastic gradient descent. In *SIGKDD International Conference on Knowledge Discovery and Data Mining*, pages 69–77, 2011. DOI: 10.1145/2020408.2020426. 17

James E. Gentle. *Matrix Algebra: Theory, Computations, and Applications in Statistics*. Springer Texts in Statistics, Springer, 2007. DOI: 10.1007/978-0-387-70873-7. 17

Ian Goodfellow, Yoshua Bengio, and Aaron Courville. *Deep Learning*, chapter 14. MIT Press, 2016. http://www.deeplearningbook.org/contents/autoencoders.html 113

Ian J Goodfellow, Jonathon Shlens, and Christian Szegedy. Explaining and harnessing adversarial examples. In *International Conference on Learning Representations*, 2015. 68, 116, 119, 128, 129

Kathrin Grosse, Nicolas Papernot, Praveen Manoharan, Michael Backes, and Patrick McDaniel. Adversarial perturbations against deep neural networks for malware classification. In *European Symposium on Research in Computer Security*, 2017. 74

Michael Grosshans, Christoph Sawade, Michael Brückner, and Tobias Scheffer. Bayesian games for adversarial regression problems. In *International Conference on International Conference on Machine Learning*, pages 55–63, 2013. 42, 51, 73, 75, 132

Claudio Guarnieri, Alessandro Tanasi, Jurriaan Bremer, and Mark Schloesser. Cuckoo sandbox: A malware analysis system, 2012. http://www.cuckoosandbox.org/ 30

Carlos Guestrin, Daphne Koller, Ronald Parr, and Shobha Venkataraman. Efficient solution algorithms for factored MDPS. *Journal of Artificial Intelligence Research*, 19:399–468, 2003. 17

Chen Hajaj and Yevgeniy Vorobeychik. Adversarial task assignment. In *International Joint Conference on Artificial Intelligence*, to appear, 2018. 132

S. Hanna, L. Huang, E. Wu, S. Li, C. Chen, and D. Song. Juxtapp: A scalable system for detecting code reuse among android applications. In *International Conference on Detection of Intrusions and Malware, and Vulnerability Assessment*, pages 62–81, 2013. DOI: 10.1007/978-3-642-37300-8_4. 13

Moritz Hardt, Nimrod Megiddo, Christos Papadimitriou, and Mary Wootters. Strategic classification. In *Proc. of the ACM Conference on Innovations in Theoretical Computer Science*, pages 111–122, 2016. DOI: 10.1145/2840728.2840730. 50

Trevor Hastie, Robert Tibshirani, and Jerome Friedman. *The Elements of Statistical Learning: Data Mining, Inference, and Prediction*, 2nd ed. Springer Series in Statistics, Springer, 2016. DOI: 10.1007/978-0-387-84858-7. 16

Klaus-U. Hoffgen, Hans-U. Simon, and Kevin S. Van Horn. Robust trainability of single neurons. *Journal of Computer and System Sciences*, 50(1):114–125, 1995. DOI: 10.1006/jcss.1995.1011. 48

Holger H. Hoos and Thomas Stützle. *Stochastic Local Search: Foundations and Applications*. The Morgan Kaufmann Series in Artificial Intelligence, Morgan Kaufmann, 2004. DOI: 10.1016/B978-1-55860-872-6.X5016-1. 40

Harold Hotelling. Analysis of a complex of statistical variables into principal components. *Journal of Educational Psychology*, 24(6):417, 1933. DOI: 10.1037/h0070888. 107

Matthew Jagielski, Alina Oprea, Battista Biggio, Chang Liu, Cristina Nita-Rotaru, and Bo Li. Manipulating machine learning: Poisoning attacks and countermeasures for regression learning. In *IEEE Symposium on Security and Privacy*, 2018. 98

Prateek Jain, Praneeth Netrapalli, and Sujay Sanghavi. Low-rank matrix completion using alternating minimization. In *STOC*, 2013. DOI: 10.1145/2488608.2488693. 12

Ian Jolliffe. *Principal Component Analysis*. Wiley Online Library, 2002. DOI: 10.1002/9781118445112.stat06472. 107

Ian T. Jolliffe. A note on the use of principal components in regression. *Applied Statistics*, pages 300–303, 1982. DOI: 10.2307/2348005. 105

Jeff Kahn, Gil Kalai, and Nathan Linial. The influence of variables on Boolean functions. In *Foundations of Computer Science, 29th Annual Symposium on*, pages 68–80, IEEE, 1988. DOI: 10.1109/sfcs.1988.21923. 72

Adam Kalai, Adam R. Klivans, Yishai Mansour, and Rocco A. Servedio. Agnostically learning halfspaces. *SIAM Journal on Computing*, 37(6):1777–1805, 2008. DOI: 10.1137/060649057. 110

Murat Kantarcioglu, Bowei Xi, and Chris Clifton. Classifier evaluation and attribute selection against active adversaries. *Data Mining and Knowledge Discovery*, 22(1–2):291–335, 2011. https://doi.org/10.1007/s10618--010-0197-3 DOI: 10.1007/s10618-010-0197-3. 74

Alex Kantchelian, J. D. Tygar, and Anthony D. Joseph. Evasion and hardening of tree ensemble classifiers. In *International Conference on Machine Learning*, 2016. 74

Michael Kearns and Ming Li. Learning in the presence of malicious errors. *SIAM Journal on Computing*, 22(4):807–837, 1993. DOI: 10.1137/0222052. 96, 99, 110

Hans Kellerer, Ulrich Pferschy, and David Pisinger. *Knapsack Problems*. Springer, 2004. DOI: 10.1007/978-3-540-24777-7. 37

Adam R. Klivans, Philip M. Long, and Rocco A. Servedio. Learning halfspaces with malicious noise. *Journal of Machine Learning Research*, 10:2715–2740, 2009. DOI: 10.1007/978-3-642-02927-1_51. 99, 100, 101, 110

Marius Kloft and Pavel Laskov. Security analysis of online centroid anomaly detection. *Journal of Machine Learning Research*, 13:3681–3724, 2012. 13, 16, 86, 97

Pang Wei Koh and Percy Liang. Understanding black-box predictions via influence functions. In *International Conference on Machine Learning*, 2017. 97

Alexey Kurakin, Ian J. Goodfellow, and Samy Bengio. Adversarial examples in the physical world. *CoRR*, abs/1607.02533, 2016. http://arxiv.org/abs/1607.02533 128

A. Lakhina, M. Crovella, and C. Diot. Diagnosing network-wide traffic anomalies. In *SIG-COMM Conference*, 2004. DOI: 10.1145/1030194.1015492. 14, 16

Bertrand Lebichot, Fabian Braun, Olivier Caelen, and Marco Saerens. A graph-based, semi-supervised, credit card fraud detection system. In *International Workshop on Complex Networks and their Applications*, 2016. DOI: 10.1007/978-3-319-50901-3_57. 10

Bo Li and Yevgeniy Vorobeychik. Feature cross-substitution in adversarial classification. In *Neural Information Processing Systems*, pages 2087–2095, 2014. 50, 74

Bo Li and Yevgeniy Vorobeychik. Scalable optimization of randomized operational decisions in adversarial classification settings. In *Conference on Artificial Intelligence and Statistics*, 2015. 69, 72, 73, 75

Bo Li and Yevgeniy Vorobeychik. Evasion-robust classification on binary domains. *ACM Transactions on Knowledge Discovery from Data*, 12(4):Article 50, 2018. DOI: 10.1145/3186282. 56, 59, 68, 74, 125

Bo Li, Yining Wang, Aarti Singh, and Yevgeniy Vorobeychik. Data poisoning attacks on factorization-based collaborative filtering. In *Neural Information Processing Systems*, pages 1885–1893, 2016. 17, 87, 89, 91, 97

Chang Liu, Bo Li, Yevgeniy Vorobeychik, and Alina Oprea. Robust linear regression against training data poisoning. In *Workshop on Artificial Intelligence and Security*, 2017. DOI: 10.1145/3128572.3140447. 99, 105, 106, 110, 111

Daniel Lowd and Christopher Meek. Adversarial learning. In *ACM SIGKDD International Conference on Knowledge Discovery in Data Mining*, pages 641–647, 2005a. DOI: 10.1145/1081870.1081950. 25, 37, 48, 50, 51

Daniel Lowd and Christopher Meek. Good word attacks on statistical spam filters. In *Conference on Email and Anti-Spam*, 2005b. 3

Chunchuan Lyu, Kaizhu Huang, and Hai-Ning Liang. A unified gradient regularization family for adversarial examples. In *IEEE International Conference on Data Mining*, pages 301–309, 2015. DOI: 10.1109/icdm.2015.84. 119, 128, 129

Aleksander Madry, Aleksandar Makelov, Ludwig Schmidt, Dimitris Tsipras, and Adrian Vladu. Towards deep learning models resistant to adversarial attacks. In *International Conference on Learning Representations*, 2018. 120, 125, 129

S. Martello and P. Toth. *Knapsack Problems: Algorithms and Computer Implementations*. John Wiley & Sons, 1990. 37

Garth P. McCormick. Computability of global solutions to factorable nonconvex programs: Part i—convex underestimating problems. *Mathematical Programming*, 10(1):147–175, 1976. DOI: 10.1007/bf01580665. 58

Shike Mei and Xiaojin Zhu. Using machine teaching to identify optimal training-set attacks on machine learners. In *AAAI Conference on Artificial Intelligence*, pages 2871–2877, 2015a. 93, 94, 97

Shike Mei and Xiaojin Zhu. The security of latent Dirichlet allocation. In *International Conference on Artificial Intelligence and Statistics*, pages 681–689, 2015b. 97

German E. Melo-Acosta, Freddy Duitama-Munoz, and Julian D. Arias-Londono. Fraud detection in big data using supervised and semi-supervised learning techniques. In *IEEE Colombian Conference on Communications and Computing*, 2017. DOI: 10.1109/colcom-con.2017.8088206. 10

John D. Montgomery. Spoofing, market manipulation, and the limit-order book. *Technical Report*, Navigant Economics, 2016. http://ssrn.com/abstract=2780579 DOI: 10.2139/ssrn.2780579. 3

Seyed-Mohsen Moosavi-Dezfooli, Alhussein Fawzi, and Pascal Frossard. DeepFool: A simple and accurate method to fool deep neural networks. In *Conference on Computer Vision and Pattern Recognition*, pages 2574–2582, 2016a. DOI: 10.1109/cvpr.2016.282. 117

Seyed-Mohsen Moosavi-Dezfooli, Alhussein Fawzi, and Pascal Frossard. Deepfool: A simple and accurate method to fool deep neural networks. In *IEEE Conference on Computer Vision and Pattern Recognition*, pages 2574–2582, 2016b. DOI: 10.1109/cvpr.2016.282. 128

Seyed-Mohsen Moosavi-Dezfooli, Alhussein Fawzi, Omar Fawzi, and Pascal Frossard. Universal adversarial perturbations. In *IEEE Conference on Computer Vision and Pattern Recognition*, pages 1765–1773, 2017. DOI: 10.1109/cvpr.2017.17. 128

Nagarajan Natarajan, Inderjit S. Dhillon, Pradeep Ravikumar, and Ambuj Tewari. Learning with noisy labels. In *Proc. of the 26th International Conference on Neural Information Processing Systems*, vol. 1, pages 1196–1204, 2013. 96

Blaine Nelson, Benjamin I. P. Rubinstein, Ling Huang, Anthony D. Joseph, and J. D. Tygar. Classifier evasion: Models and open problems. In *Privacy and Security Issues in Data Mining and Machine Learning—International ECML/PKDD Workshop*, pages 92–98, 2010. DOI: 10.1007/978-3-642-19896-0_8. 50

Blaine Nelson, Benjamin I. P. Rubinstein, Ling Huang, Anthony D. Joseph, Steven J. Lee, Satish Rao, and J. D. Tygar. Query strategies for evading convex-inducing classifiers. *Journal of Machine Learning Research*, pages 1293–1332, 2012. 25, 48, 50

Jorge Nocedal and Stephen Wright. *Numerical Optimization*, 2nd ed. Springer Series in Operations Research and Financial Engineering, Springer, 2006. DOI: 10.1007/b98874. 36, 40

Ryan O'Donnell. Some topics in analysis of Boolean functions. In *ACM Symposium on Theory of Computing*, pages 569–578, 2008. DOI: 10.1145/1374376.1374458. 72

N. Papernot, P. McDaniel, X. Wu, S. Jha, and A. Swami. Distillation as a defense to adversarial perturbations against deep neural networks. In *IEEE Symposium on Security and Privacy*, pages 582–597, 2016a. DOI: 10.1109/sp.2016.41. 127, 129

Nicolas Papernot and Patrick McDaniel. Deep k-nearest neighbors: Towards confident, interpretable and robust deep learning. *Arxiv Preprint, ArXiv:1803.04765*, 2018. 129, 133

Nicolas Papernot, Patrick McDaniel, Somesh Jha, Matt Fredrikson, Z. Berkay Celik, and Ananthram Swami. The limitations of deep learning in adversarial settings. In *IEEE European Symposium on Security and Privacy*, 2016b. DOI: 10.1109/eurosp.2016.36. 121

Nicolas Papernot, Patrick D. McDaniel, and Ian J. Goodfellow. Transferability in machine learning: From phenomena to black-box attacks using adversarial samples. *Arxiv*, preprint, 2016c. 50, 123, 128

Nicolas Papernot, Patrick McDaniel, Ian Goodfellow, Somesh Jha, Z. Berkay Celik, and Ananthram Swami. Practical black-box attacks against machine learning. In *ACM Asia Conference on Computer and Communications Security*, pages 506–519, 2017. DOI: 10.1145/3052973.3053009. 123, 128

R. Perdisci, D. Ariu, and G. Giacinto. Scalable fine-grained behavioral clustering of http-based malware. *Computer Networks*, 57(2):487–500, 2013. DOI: 10.1016/j.comnet.2012.06.022. 13

Aditi Raghunathan, Jacob Steinhardt, and Percy Liang. Certified defenses against adversarial examples. In *International Conference on Learning Representations*, 2018. 120, 126, 129

Anand Rajaraman and Jeffrey David Ullman. *Mining of Massive Datasets*. Cambridge University Press, 2012. DOI: 10.1017/cbo9781139058452. 1

Bita Darvish Rouhani, Mohammad Samragh, Tara Javidi, and Farinaz Koushanfar. Curtail: Characterizing and thwarting adversarial deep learning. *Arxiv Preprint, ArXiv:1709.02538*, 2017. 129

Benjamin I. P. Rubinstein, Blaine Nelson, Ling Huang, Anthony D. Joseph, Shing hon Lau, Satish Rao, Nina Taft, and J. D. Tygar. ANTIDOTE: Understanding and defending against poisoning of anomaly detectors. In *Internet Measurement Conference*, 2009. DOI: 10.1145/1644893.1644895. 86, 97

Paolo Russu, Ambra Demontis, Battista Biggio, Giorgio Fumera, and Fabio Roli. Secure kernel machines against evasion attacks. In *Proc. of the ACM Workshop on Artificial Intelligence and Security*, pages 59–69, 2016. DOI: 10.1145/2996758.2996771. 67, 75

Rocco A. Servedio. Smooth boosting and learning with malicious noise. *Journal of Machine Learning Research*, 4:633–648, 2003. DOI: 10.1007/3-540-44581-1_31. 110

Mahmood Sharif, Sruti Bhagavatula, Lujo Bauer, and Michael K. Reiter. Accessorize to a crime: Real and stealthy attacks on state-of-the-art face recognition. In *ACM*

SIGSAC Conference on Computer and Communications Security, pages 1528–1540, 2016. DOI: 10.1145/2976749.2978392. 122, 128

Andrew Smith, Jian Lou, and Yevgeniy Vorobeychik. Multidefender security games. *IEEE Intelligent Systems*, 32(1):50–60, 2017. 134

C. Smutz and A. Stavrou. Malicious PDF detection using metadata and structural features. In *Annual Computer Security Applications Conference*, pages 239–248, 2012. DOI: 10.1145/2420950.2420987. 10, 29

Suvrit Sra and Inderjit S. Dhillon. Generalized nonnegative matrix approximations with Bregman divergences. In *Neural Information Processing Systems*, pages 283–290, 2006. 17

N. Šrndic and P. Laskov. Practical evasion of a learning-based classifier: A case study. In *IEEE Symposium on Security and Privacy*, pages 197–211, 2014. DOI: 10.1109/sp.2014.20. 25, 29

Nedim Šrndić and Pavel Laskov. Hidost: A static machine-learning-based detector of malicious files. *EURASIP Journal on Information Security*, (1):22, 2016. DOI: 10.1186/s13635-016-0045-0. 10

Robert St. Aubin, Jesse Hoey, and Craig Boutilier. Apricodd: Approximate policy construction using decision diagrams. In *NIPS*, pages 1089–1095, 2000. 17

Jacob Steinhardt, Pang Wei Koh, and Percy Liang. Certified defenses for data poisoning attacks. In *Neural Information Processing Systems*, 2017. 111

David Stevens and Daniel Lowd. On the hardness of evading combinations of linear classifiers. In *ACM Workshop on Artificial Intelligence and Security*, 2013. DOI: 10.1145/2517312.2517318. 134

Octavian Suciu, Radu Marginean, Yigitcan Kaya, Hal Daume III, and Tudor Dumitras. When does machine learning FAIL? Generalized transferability for evasion and poisoning attacks. In *USENIX Security Symposium*, 2018. 25, 98

Richard S. Sutton and Andrew G. Barto. *Reinforcement Learning: An Introduction*. Adaptive Computation and Machine Learning, A Bradford Book, 1998. 14, 16

Christian Szegedy, Wojciech Zaremba, Ilya Sutskever, Joan Bruna, Dumitru Erhan, Ian J. Goodfellow, and Rob Fergus. Intriguing properties of neural networks. In *International Conference on Learning Representations*, 2013. 116, 117, 123, 127, 128, 129

Milind Tambe, Ed. *Security and Game Theory: Algorithms, Deployed Systems, Lessons Learned*. Cambridge University Press, 2011. DOI: 10.1017/cbo9780511973031. 56, 69, 75, 133

Acar Tamersoy, Kevin Roundy, and Duen Horng Chau. Guilt by association: Large scale malware detection by mining file-relation graphs. In *SIGKDD International Conference on Knowledge Discovery and Data Mining*, 2014. DOI: 10.1145/2623330.2623342. 10

Choon Hai Teo, Amir Globerson, Sam Roweis, and Alexander J. Smola. Convex learning with invariances. In *Neural Information Processing Systems*, 2007. 66, 74, 129

Liang Tong, Bo Li, Chen Hajaj, Chaowei Xiao, and Yevgeniy Vorobeychik. A framework for validating models of evasion a acks on machine learning, with application to PDF malware detection. *Arxiv Preprint, ArXiv:1708.08327v3*, 2018a. 134

Liang Tong, Sixie Yu, Scott Alfeld, and Yevgeniy Vorobeychik. Adversarial regression with multiple learners. In *International Conference on Machine Learning*, to appear, 2018b. 134

Leslie Valiant. Learning disjunctions of conjunctions. In *International Joint Conference on Artificial Intelligence*, pages 560–566, 1985. 110

Vladimir Vapnik. *The Nature of Statistical Learning Theory*, 2nd ed. Information Science and Statistics, Springer, 1999. DOI: 10.1007/978-1-4757-3264-1. 16

Yevgeniy Vorobeychik and Bo Li. Optimal randomized classification in adversarial settings. In *International Conference on Autonomous Agents and Multiagent Systems*, pages 485–492, 2014. 51

Vladimir Vovk, Alex Gammerman, and Glenn Shafer. *Algorithmic learning in a random world*. Springer Verlag, 2005. 133

Gang Wang, Tianyi Wang, Haitao Zheng, and Ben Y. Zhao. Man vs. machine: Practical adversarial detection of malicious crowdsourcing workers. In *USENIX Security Symposium*, pages 239–254, 2014. 132

Ke Wang, Janak J. Parekh, and Salvatore J. Stolfo. Anagram: A content anomaly detector resistant to mimicry attack. In *Recent Advances in Intrusion Detection*, pages 226–248, 2006. DOI: 10.1007/11856214_12. 14

Max Welling and Yee W. Teh. Bayesian learning via stochastic gradient Langevin dynamics. In *Proc. of the 28th International Conference on Machine Learning (ICML-11)*, pages 681–688, 2011. 93

Eric Wong and J. Zico Kolter. Provable defenses against adversarial examples via the convex outer adversarial polytope. In *International Conference on Machine Learning*, 2018. 120, 126, 129

Chaowei Xiao, Jun-Yan Zhu, Bo Li, Warren He, Mingyan Liu, and Dawn Song. Spatially transformed adversarial examples. In *International Conference on Learning Representations*, 2018. 134

Han Xiao, Huang Xiao, and Claudia Eckert. Adversarial label flips attack on support vector machines. In *European Conference on Artificial Intelligence*, 2012. DOI: 10.3233/978-1-61499-098-7-870. 80, 96

Huang Xiao, Battista Biggio, Blaine Nelson, HanXiao, Claudia Eckert, and Fabio Roli. Support vector machines under adversarial label contamination. *Neurocomputing*, 160:53–62, 2015. DOI: 10.1016/j.neucom.2014.08.081. 96

Huan Xu, Constantine Caramanis, and Shie Mannor. Robust regression and lasso. In *Neural Information Processing Systems 21*, pages 1801–1808, 2009a. DOI: 10.1109/tit.2010.2048503. 111

Huan Xu, Constantine Caramanis, and Shie Mannor. Robustness and regularization of support vector machines. *Journal of Machine Learning Research*, 10:1485–1510, 2009b. 67, 75, 129

Huan Xu, Constantin Caramanis, and Sujay Sanghavi. Robust PCA via outlier pursuit. *IEEE Transactions on Information Theory*, 58(5):3047–3064, 2012. DOI: 10.1109/tit.2011.2173156. 111

Huan Xu, Constantin Caramanis, and Shie Mannor. Outlier-robust PCA: The high-dimensional case. *IEEE Transactions on Information Theory*, 59(1):546–572, 2013. DOI: 10.1109/tit.2012.2212415. 111

Weilin Xu, Yanjun Qi, and David Evans. Automatically evading classifiers: A case study on PDF malware classifiers. In *Network and Distributed System Security Symposium*, 2016. DOI: 10.14722/ndss.2016.23115. 30, 47, 50

Yanfang Ye, Tao Li, Donald Adjeroh, and S. Sitharama Iyengar. A survey on malware detection using data mining techniques. *ACM Computing Surveys*, 50(3), 2017. DOI: 10.1145/3073559. 10

F. Zhang, P. P. K. Chan, B. Biggio, D. S. Yeung, and F. Roli. Adversarial feature selection against evasion attacks. *IEEE Transactions on Cybernetics*, 2015. DOI: 10.1109/tcyb.2015.2415032. 50

Yan Zhou and Murat Kantarcioglu. Modeling adversarial learning as nested stackelberg games. In *Advances in Knowledge Discovery and Data Mining—20th Pacific-Asia Conference, PAKDD, Proceedings, Part II*, pages 350–362, Auckland, New Zealand, April 19–22, 2016. https://doi.org/10.1007/978--3-319-31750-2_28 DOI: 10.1007/978-3-319-31750-2_28. 75

Yan Zhou, Murat Kantarcioglu, Bhavani M. Thuraisingham, and Bowei Xi. Adversarial support vector machine learning. In *ACM SIGKDD International Conference on Knowledge Discovery and Data Mining*, pages 1059–1067, 2012. DOI: 10.1145/2339530.2339697. 51, 74

Authors' Biographies

YEVGENIY VOROBEYCHIK

Yevgeniy Vorobeychik is an Assistant Professor of Computer Science, Computer Engineering, and Biomedical Informatics at Vanderbilt University. Previously, he was a Principal Research Scientist at Sandia National Laboratories. Between 2008 and 2010, he was a post-doctoral research associate at the University of Pennsylvania Computer and Information Science department. He received Ph.D. (2008) and M.S.E. (2004) degrees in Computer Science and Engineering from the University of Michigan, and a B.S. degree in Computer Engineering from Northwestern University. His work focuses on game theoretic modeling of security and privacy, adversarial machine learning, algorithmic and behavioral game theory and incentive design, optimization, agent-based modeling, complex systems, network science, and epidemic control. Dr. Vorobeychik received an NSF CAREER award in 2017, and was invited to give an IJCAI-16 early career spotlight talk. He was nominated for the 2008 ACM Doctoral Dissertation Award and received honorable mention for the 2008 IFAAMAS Distinguished Dissertation Award.

MURAT KANTARCIOGLU

Murat Kantarcioglu is a Professor of Computer Science and Director of the UTD Data Security and Privacy Lab at The University of Texas at Dallas. Currently, he is also a visiting scholar at Harvard's Data Privacy Lab. He holds a B.S. in Computer Engineering from Middle East Technical University, and M.S. and Ph.D. degrees in Computer Science from Purdue University.

Dr. Kantarcioglu's research focuses on creating technologies that can efficiently extract useful information from any data without sacrificing privacy or security. His research has been supported by awards from NSF, AFOSR, ONR, NSA, and NIH. He has published over 175 peer-reviewed papers. His work has been covered by media outlets such as The Boston Globe and ABC News, among others, and has received three best paper awards. He is also the recipient of various awards including NSF CAREER award, a Purdue CERIAS Diamond Award for academic excellence, the AMIA (American Medical Informatics Association) 2014 Homer R. Warner Award, and the IEEE ISI (Intelligence and Security Informatics) 2017 Technical Achievement Award presented jointly by IEEE SMC and IEEE ITS societies for his research in data security and privacy. He is also a Distinguished Scientist of ACM.

Index

Printed in the United States
by Baker & Taylor Publisher Services